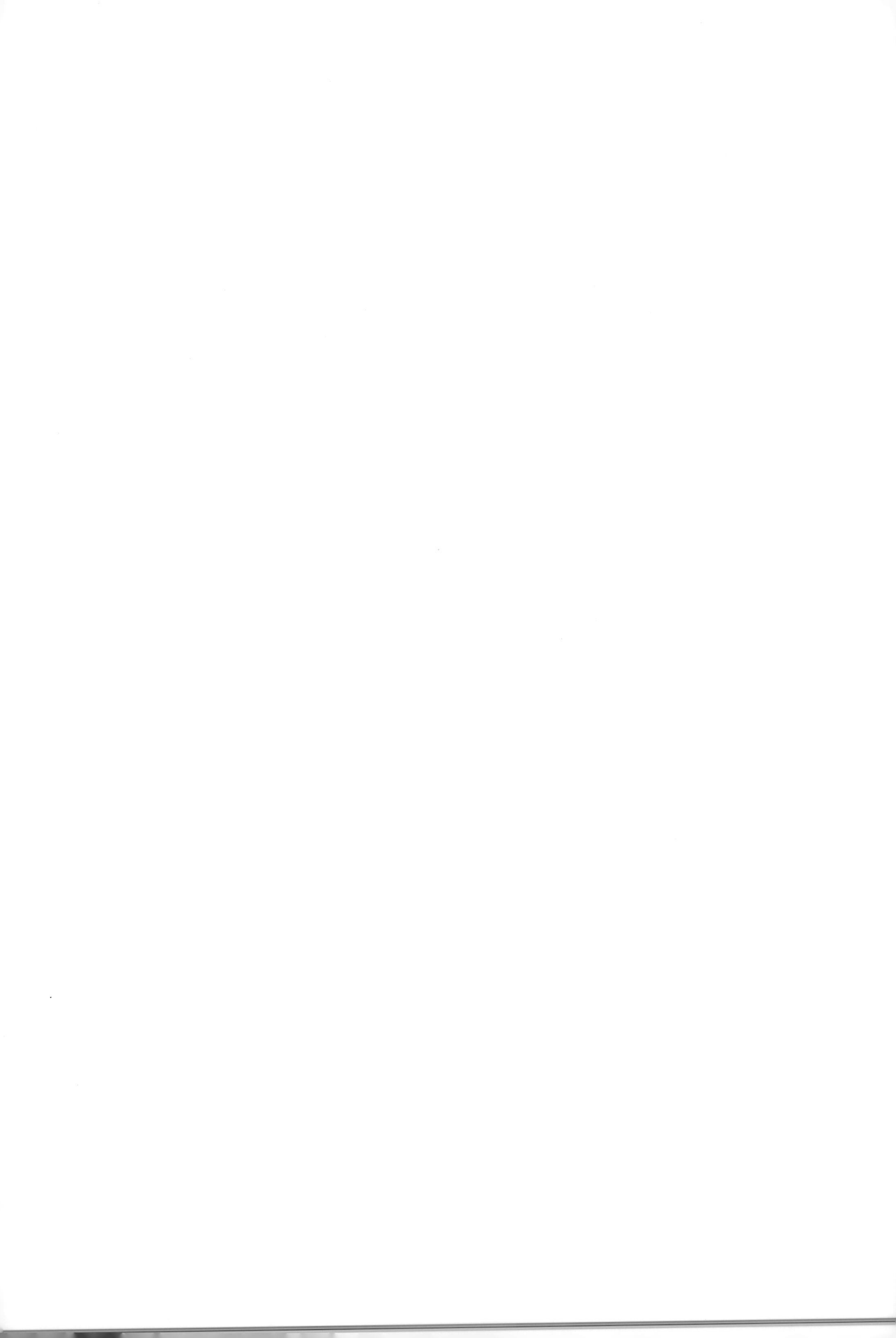

Integrated Principles
of Zoology

Integrated Principles of Zoology

Edited by
Will Fisher

🖙 Larsen & Keller
www.larsen-keller.com

Integrated Principles of Zoology
Edited by Will Fisher
ISBN: 978-1-63549-039-8 (Hardback)

© 2017 Larsen & Keller

⊟ Larsen & Keller

Published by Larsen and Keller Education,
5 Penn Plaza,
19th Floor,
New York, NY 10001, USA

Cataloging-in-Publication Data

Integrated principles of zoology / edited by Will Fisher.
    p. cm.
Includes bibliographical references and index.
ISBN 978-1-63549-039-8
1. Zoology. 2. Animals. 3. Biology. I. Fisher, Will.
QL45.2 .I58 2017
590--dc23

The publisher's policy is to use permanent paper from mills that operate a sustainable forestry policy. Furthermore, the publisher ensures that the text paper and cover boards used have met acceptable environmental accreditation standards.

Printed and bound in the United States of America.

For more information regarding Larsen and Keller Education and its products, please visit the publisher's website www.larsen-keller.com

Table of Contents

Permissions

Index

Preface

The book aims to shed light on some of the fundamental aspects of zoology. It talks about the various techniques used in this study and its different applications. Zoology, as a part of biology, studies the evolution, classification, embryology, structure, and behavior of all living and extinct animals. It also lays focus on the interaction between the animals and their environment. This book is a compilation of chapters that discuss the most vital concepts in the field of zoology. This textbook, with its detailed analyses and data, will prove immensely beneficial to graduates and post-graduates involved in this area. Those with an interest in this field would find this text helpful.

To facilitate a deeper understanding of the contents of this book a short introduction of every chapter is written below:

Chapter 1- Zoology is an overarching area of study that seeks to analyse the relationship between animals and their environment. The natural sciences, the forerunner of zoology, established the concept of the symbiotic relationship between animals, their dietary and social behavior and their habitat. Scientific progress has classified the natural sciences into ethology, biogeography and so on. This chapter will provide an integrated understanding of zoology.

Chapter 2- Darwinian theories of evolution of species was a landmark in the study of zoology. Along with a systematized taxonomy of species classification, Darwin's theories paved the way for the theories of natural selection and the concept of the survival of the fittest. This, in turn, altered the perception of human social behavior, culture and savagery as well as life in the modern world.

Chapter 3- The progress that the natural sciences has made in the last hundred and fifty years has no parallel. Such a radical change in the sciences has created new ways of inference and deduction. The chapter strategically encompasses and incorporates the major components and key concepts of zoology, providing a complete understanding.

Chapter 4- Animals that possess backbones and chordates are known as vertebrates and those without them are invertebrates. Vertebrates evolved from invertebrate species in the Cambrian and Devonian period of the Earth's history. Binomial classification, or naming of species relies on such anatomical differentiation among animals. The basic classification of zoology is dealt with in this chapter.

Chapter 5- Reproduction is the activity of creating newer individuals of the same species for species survival and genetic propagation. Two types of reproduction exist in animals: sexual reproduction and asexual reproduction. This chapter deals with these themes as well as related ones such as sexual dimorphism and sexual mimicry. Zoology is best understood in confluence with the major topics listed in the following chapter.

Chapter 6- Zoological sciences contain various theories about the natural world. Such theories are derived from phenomena that can be observed among animals in the wild. These theories seek to expand the scope of zoology while being faithful to its basic principles. Some themes explored in this chapter include alpha behavior among male members of species, bipedalism and the autotomy of limbs.

Finally, I would like to thank the entire team involved in the inception of this book for their valuable time and contribution. This book would not have been possible without their efforts. I would also like to thank my friends and family for their constant support.

Editor

Introduction to Zoology

Zoology is an overarching area of study that seeks to analyse the relationship between animals and their environment. The natural sciences, the forerunner of zoology, established the concept of the symbiotic relationship between animals, their dietary and social behaviour and their habitat. Scientific progress has classified the natural sciences into ethology, biogeography and so on. This chapter will provide an integrated understanding of zoology.

Zoology or animal biology is the branch of biology that relates to the animal kingdom, including the structure, embryology, evolution, classification, habits, and distribution of all animals, both living and extinct, and how they interact with their ecosystems.

History

Ancient History to Darwin

Conrad Gesner (1516–1565). His *Historiae animalium* is considered the beginning of modern zoology.

The history of zoology traces the study of the animal kingdom from ancient to modern times. Although the concept of *zoology* as a single coherent field arose much later, the zoological sciences emerged from natural history reaching back to the works of Aristotle and Galen in the ancient Greco-Roman world. This ancient work was further developed in the Middle Ages by Muslim physicians and scholars such as Albertus Magnus. During the Renaissance and early modern period, zoological thought was revolutionized in Europe by a renewed interest in empiricism and the discovery of many novel organisms. Prominent in this movement were Vesalius and William Harvey, who used experimentation and careful observation in physiology, and naturalists such as Carl Linnaeus and Buffon who began to classify the diversity of life and the fossil record, as well as the development and behavior of organisms. Microscopy revealed the previously unknown world of microorganisms, laying the groundwork for cell theory. The growing importance of natural theology, partly a response to the rise of mechanical philosophy, encouraged the growth of natural history (although it entrenched the argument from design).

Over the 18th and 19th centuries, zoology became an increasingly professional scientific discipline. Explorer-naturalists such as Alexander von Humboldt investigated the interaction between organisms and their environment, and the ways this relationship depends on geography, laying the foundations for biogeography, ecology and ethology. Naturalists began to reject essentialism and consider the importance of extinction and the mutability of species. Cell theory provided a new perspective on the fundamental basis of life.

Post-darwin

These developments, as well as the results from embryology and paleontology, were synthesized in Charles Darwin's theory of evolution by natural selection. In 1859, Darwin placed the theory of organic evolution on a new footing, by his discovery of a process by which organic evolution can occur, and provided observational evidence that it had done so.

Darwin gave new direction to morphology and physiology, by uniting them in a common biological theory: the theory of organic evolution. The result was a reconstruction of the classification of animals upon a genealogical basis, fresh investigation of the development of animals, and early attempts to determine their genetic relationships. The end of the 19th century saw the fall of spontaneous generation and the rise of the germ theory of disease, though the mechanism of inheritance remained a mystery. In the early 20th century, the rediscovery of Mendel's work led to the rapid development of genetics by Thomas Hunt Morgan and his students, and by the 1930s the combination of population genetics and natural selection in the "neo-Darwinian synthesis".

Research

Structural

Cell biology studies the structural and physiological properties of cells, including their behavior, interactions, and environment. This is done on both the microscopic and molecular levels, for single-celled organisms such as bacteria as well as the specialized cells in multicellular organisms such as humans. Understanding the structure and function of cells is fundamental to all of the biological sciences. The similarities and differences between cell types are particularly relevant to molecular biology.

Anatomy considers the forms of macroscopic structures such as organs and organ systems. It focuses on how organs and organ systems work together in the bodies of humans and animals, in addition to how they work independently. Anatomy and cell biology are two studies that are closely related, and can be categorized under "structural" studies.

Physiological

Physiology studies the mechanical, physical, and biochemical processes of living organisms by attempting to understand how all of the structures function as a whole. The theme of "structure to function" is central to biology. Physiological studies have traditionally been divided into plant physiology and animal physiology, but some principles of physiology are universal, no matter what particular organism is being studied. For example, what is learned about the physiology of yeast cells can also apply to human cells. The field of animal physiology extends the tools and methods of human physiology to non-human species. Physiology studies how for example nervous, immune, endocrine, respiratory, and circulatory systems, function and interact.

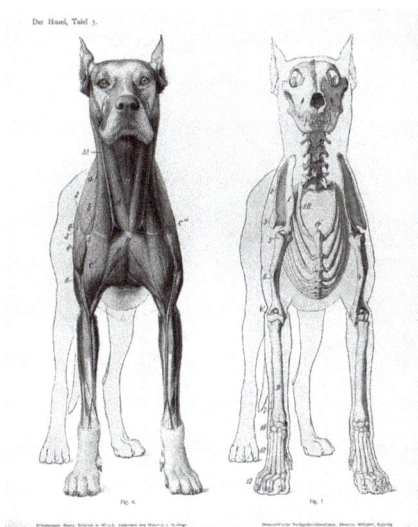

Animal anatomical engraving from *Handbuch der Anatomie der Tiere für Künstler*.

Evolutionary

Evolutionary research is concerned with the origin and descent of species, as well as their change over time, and includes scientists from many taxonomically oriented disciplines. For example, it generally involves scientists who have special training in particular organisms such as mammalogy, ornithology, herpetology, or entomology, but use those organisms as systems to answer general questions about evolution.

Evolutionary biology is partly based on paleontology, which uses the fossil record to answer questions about the mode and tempo of evolution, and partly on the developments in areas such as population genetics and evolutionary theory. Following the development of DNA fingerprinting techniques in the late 20th century, the application of these techniques in zoology has increased the understanding of animal populations. In the 1980s, developmental biology re-entered evolutionary biology from its initial exclusion from the modern synthesis through the study of evolutionary developmental biology. Related fields often considered part of evolutionary biology are phylogenetics, systematics, and taxonomy.

Classification

Scientific classification in zoology, is a method by which zoologists group and categorize organisms by biological type, such as genus or species. Biological classification is a form of scientific taxonomy. Modern biological classification has its root in the work of Carl Linnaeus, who grouped species according to shared physical characteristics. These groupings have since been revised to improve consistency with the Darwinian principle of common descent. Molecular phylogenetics, which uses DNA sequences as data, has driven many recent revisions and is likely to continue to do so. Biological classification belongs to the science of zoological systematics.

Linnaeus's table of the animal kingdom from the first edition of *Systema Naturae* (1735).

Many scientists now consider the five-kingdom system outdated. Modern alternative classification systems generally start with the three-domain system: Archaea (originally Archaebacteria); Bacteria (originally Eubacteria); Eukaryota (including protists, fungi, plants, and animals) These domains reflect whether the cells have nuclei or not, as well as differences in the chemical composition of the cell exteriors.

Further, each kingdom is broken down recursively until each species is separately classified. The order is: Domain; kingdom; phylum; class; order; family; genus; species. The scientific name of an organism is generated from its genus and species. For example, humans are listed as *Homo sapiens. Homo* is the genus, and *sapiens* the specific epithet, both of them combined make up the species name. When writing the scientific name of an organism, it is proper to capitalize the first letter in the genus and put all of the specific epithet in lowercase. Additionally, the entire term may be italicized or underlined.

The dominant classification system is called the Linnaean taxonomy. It includes ranks and binomial nomenclature. The classification, taxonomy, and nomenclature of zoological organisms is administered by the International Code of Zoological Nomenclature, and International Code of Nomenclature of Bacteria for animals and bacteria, respectively. The classification of viruses, viroids, prions, and all other sub-viral agents that demonstrate biological characteristics is conducted by the International Code of Virus classification and nomenclature. However, several other viral classification systems do exist.

A merging draft, BioCode, was published in 1997 in an attempt to standardize nomenclature in these areas, but has yet to be formally adopted. The BioCode draft has received little attention since 1997; its originally planned implementation date of January 1, 2000, has passed unnoticed.

However, a 2004 paper concerning the cyanobacteria does advocate a future adoption of a Bio-Code and interim steps consisting of reducing the differences between the codes. The International Code of Virus Classification and Nomenclature (ICVCN) remains outside the BioCode.

Ethology

Kelp gull chicks peck at red spot on mother's beak to stimulate the regurgitating reflex.

Ethology is the scientific and objective study of animal behavior under natural conditions, as opposed to behaviourism, which focuses on behavioral response studies in a laboratory setting. Ethologists have been particularly concerned with the evolution of behavior and the understanding of behavior in terms of the theory of natural selection. In one sense, the first modern ethologist was Charles Darwin, whose book, *The Expression of the Emotions in Man and Animals,* influenced many future ethologists.

Biogeography

Biogeography studies the spatial distribution of organisms on the Earth, focusing on topics like plate tectonics, climate change, dispersal and migration, and cladistics. The creation of this study is widely accredited to Alfred Russel Wallace, a British biologist who had some of his work jointly published with Charles Darwin.

Branches of Zoology

Although the study of animal life is ancient, its scientific incarnation is relatively modern. This mirrors the transition from natural history to biology at the start of the 19th century. Since Hunter and Cuvier, comparative anatomical study has been associated with morphography, shaping the modern areas of zoological investigation: anatomy, physiology, histology, embryology, teratology and ethology. Modern zoology first arose in German and British universities. In Britain, Thomas Henry Huxley was a prominent figure. His ideas were centered on the morphology of animals. Many consider him the greatest comparative anatomist of the latter half of the 19th century. Similar to Hunter, his courses were composed of lectures and laboratory practical classes in contrast to the previous format of lectures only.

Gradually zoology expanded beyond Huxley's comparative anatomy to include the following sub-disciplines:

- Zoography, also known as *descriptive zoology*, describes animals and their habitats

- Comparative anatomy studies the structure of animals

- Animal physiology

- Behavioral ecology

- Ethology studies animal behavior

- Invertebrate zoology

- Vertebrate zoology

- Soil zoology

- Comparative zoology

- The various taxonomically oriented disciplines such as mammalogy, herpetology, ornithology and entomology identify and classify species and study the structures and mechanisms specific to those groups.

Related fields:

- Evolutionary biology: Development of both animals and plants is considered in the articles on evolution, population genetics, heredity, variation, Mendelism, reproduction.

- Molecular biology studies the common genetic and developmental mechanisms of animals and plants

- Palaeontology

- Systematics, cladistics, phylogenetics, phylogeography, biogeography and taxonomy classify and group species via common descent and regional associations.

References

- Paul S. Agutter & Denys N. Wheatley (2008). Thinking about Life: The History and Philosophy of Biology and Other Sciences. Springer. p. 43. ISBN 1-4020-8865-5.

- Saint Albertus Magnus (1999). On Animals: A Medieval Summa Zoologica. Johns Hopkins University Press. ISBN 0-8018-4823-7.

- Lois N. Magner (2002). A History of the Life Sciences, Revised and Expanded. CRC Press. pp. 133–144. ISBN 0-8247-0824-5.

- William Coleman (1978). "Chapter 2". Biology in the Nineteenth Century. Cambridge University Press. ISBN 0-521-29293-X.

- "Appendix: Frequently Asked Questions". Science and Creationism: a view from the National Academy of Sciences (php) (Second ed.). Washington, DC: The National Academy of Sciences. 1999. p. 28. ISBN -0-309-06406-6. Retrieved September 24, 2009.

- Vassiliki Betty Smocovitis (1996). Unifying Biology: The Evolutionary Synthesis and Evolutionary Biology. Princeton University Press. ISBN 0-691-03343-9.

- Heather Silyn-Roberts (2000). Writing for Science and Engineering: Papers, Presentation. Oxford: Butterworth-Heinemann. p. 198. ISBN 0-7506-4636-5.

- Wiley, R. H. (1981). "Social structure and individual ontogenies: problems of description, mechanism, and evolution" (PDF). Perspectives in ethology. 4: 105–133. Retrieved 21 December 2012.

- "Virus Taxonomy: 2011 Release (current)". , International Committee on Taxonomy of Viruses. Retrieved 21 December 2012.

- "Definition of ETHOLOGY". Merriam-Webster. Retrieved 30 October 2012. 2 : the scientific and objective study of animal behaviour especially under natural conditions

- Mehmet Bayrakdar (1983). "Al-Jahiz and the rise of biological evolution" (PDF). The Islamic Quarterly. 21: 149–55. Retrieved 21 December 2012.

Evolution of Zoology

Darwinian theories of evolution of species was a landmark in the study of zoology. Along with a systematized taxonomy of species classification, Darwin's theories paved the way for the theories of natural selection and the concept of the survival of the fittest. This, in turn, altered the perception of human social behaviour, culture and savagery as well as life in the modern world.

History of Zoology (Through 1859)

This Chapter considers the history of zoology up to the year 1859, when the theory of evolution by natural selection was proposed by Charles Darwin. The history of zoology traces the study of the animal kingdom from ancient to modern times. Although the concept of zoology as a single coherent field arose much later, the zoological sciences emerged from natural history reaching back to the works of Aristotle and Galen in the ancient Greco-Roman world. This ancient work was further developed in the Middle Ages by Muslim physicians and scholars such as Albertus Magnus. During the European Renaissance and early modern period, zoological thought was revolutionized in Europe by a renewed interest in empiricism and the discovery of many novel organisms. Prominent in this movement were Vesalius and William Harvey, who used experimentation and careful observation in physiology, and naturalists such as Carl Linnaeus and Buffon who began to classify the diversity of life and the fossil record, as well as the development and behavior of organisms. Microscopy revealed the previously unknown world of microorganisms, laying the groundwork for cell theory. The growing importance of natural theology, partly a response to the rise of mechanical philosophy, encouraged the growth of natural history (although it entrenched the argument from design).

Over the 18th and 19th centuries, zoology became increasingly professional scientific disciplines. Explorer-naturalists such as Alexander von Humboldt investigated the interaction between organisms and their environment, and the ways this relationship depends on geography—laying the foundations for biogeography, ecology and ethology. Naturalists began to reject essentialism and consider the importance of extinction and the mutability of species. Cell theory provided a new perspective on the fundamental basis of life. These developments, as well as the results from embryology and paleontology, were synthesized in Charles Darwin's theory of evolution by natural selection. In 1859, Darwin placed the theory of organic evolution on a new footing, by his discovery of a process by which organic evolution can occur, and provided observational evidence that it had done so.

Pre-scientific Zoology

Early Cultures

The earliest humans must have had and passed on knowledge about animals to increase their chances of survival. This may have included knowledge of human and animal anatomy and aspects of animal behavior (such as migration patterns). However, the first major turning point in zoological knowledge came with the Neolithic Revolution about 10,000 years ago. Humans domesticated livestock animals to accompany the resulting sedentary societies.

Animals in Ancient Egypt

An ancient Egyptian plows his fields with a pair of oxen, used as beasts of burden and a source of food.

The Egyptians believed that a balanced relationship between people and animals was an essential element of the cosmic order; thus humans, animals and plants were believed to be members of a single whole. Animals, both domesticated and wild, were therefore a critical source of spirituali companionship, and sustenance to the ancient Egyptians. Cattle were the most important livestock; the administration collected taxes on livestock in regular census, and the size of a herd reflected the prestige and importance of the estate or temple that owned them. In addition to cattle, the ancient Egyptians kept sheep, goats, and pigs. Poultry such as ducks, geese, and pigeons were captured in nets and bred on farms, where they were force-fed with dough to fatten them. The Nile provided a plentiful source of fish. Bees were also domesticated from at least the Old Kingdom, and they provided both honey and wax.

The ancient Egyptians used donkeys and oxen as beasts of burden, and they were responsible for plowing the fields and trampling seed into the soil. The slaughter of a fattened ox was also a central part of an offering ritual. Horses were introduced by the Hyksos in the Second Intermediate Period, and the camel, although known from the New Kingdom, was not used as a beast of burden until the Late Period. There is also evidence to suggest that elephants were briefly utilized in the Late Period, but largely abandoned due to lack of grazing land. Dogs, cats and monkeys were common family pets, while more exotic pets imported from the heart of Africa, such as lions, were reserved for royalty. Herodotus observed that the Egyptians were the only people to keep their animals with them in their houses. During the Predynastic and Late periods, the worship of the gods in their animal form was extremely popular, such as the cat goddess Bastet and the ibis god Thoth, and these animals were bred in large numbers on farms for the purpose of ritual sacrifice.

Eastern Ancient Cultures

The ancient cultures of Mesopotamia, the Indian subcontinent, and China, among others, produced renowned surgeons and students of the natural sciences such as Susruta and Zhang Zhongjing, reflecting independent sophisticated systems of natural philosophy. Taoist philosophers, such as Zhuangzi in the 4th century BC, expressed ideas related to evolution, such as denying the fixity of biological species and speculating that species had developed differing attributes in response to differing environments. The ancient Indian Ayurveda tradition independently developed the concept of three humours, resembling that of the four humours of ancient Greek medicine, though the Ayurvedic system included further complications, such as the body being composed of five elements and seven basic tissues. Ayurvedic writers also classified living things into four categories based on the method of birth (from the womb, eggs, heat & moisture, and seeds) and explained the conception of a fetus in detail. They also made considerable advances in the field of surgery, often without the use of human dissection or animal vivisection. One of the earliest Ayurvedic treatises was the *Sushruta Samhita*, attributed to Sushruta in the 6th century BC. It was also an early materia medica, describing 700 medicinal plants, 64 preparations from mineral sources, and 57 preparations based on animal sources. However, the roots of modern zoology are usually traced back to the secular tradition of ancient Greek philosophy.

Ancient Greek Traditions

The pre-Socratic philosophers asked many questions about life but produced little systematic knowledge of specifically zoological interest—though the attempts of the atomists to explain life in purely physical terms would recur periodically through the history of zoology. However, the medical theories of Hippocrates and his followers, especially humorism, had a lasting impact.

The philosopher Aristotle was the most influential scholar of the living world from classical antiquity. Though his early work in natural philosophy was speculative, Aristotle's later biological writings were more empirical, focusing on biological causation and the diversity of life. He made countless observations of nature, especially the habits and attributes of animals in the world around him, which he devoted considerable attention to categorizing. In all, Aristotle classified 540 animal species, and dissected at least 50. He believed that intellectual purposes, formal causes, guided all natural processes.

Aristotle, and nearly all Western scholars after him until the 18th century, believed that creatures were arranged in a graded scale of perfection rising from plants on up to humans: the *scala naturae* or Great Chain of Being. Pliny the Elder was also known for his knowledge of animals and nature, and was the most prolific compiler of zoological descriptions.

A few scholars in the Hellenistic period under the Ptolemies—particularly Herophilus of Chalcedon and Erasistratus of Chios—amended Aristotle's physiological work, even performing experimental dissections and vivisections. Claudius Galen became the most important authority on medicine and anatomy. Though a few ancient atomists such as Lucretius challenged the teleological Aristotelian viewpoint that all aspects of life are the result of design or purpose, teleology (and after the rise of Christianity, natural theology) would remain central to biological thought essentially until the 18th and 19th centuries. The ideas of the Greek traditions of zoology survived, but they were generally taken unquestioningly in medieval Europe.

Medieval and Islamic Knowledge

The decline of the Roman Empire led to the disappearance or destruction of much knowledge, though physicians still incorporated many aspects of the Greek tradition into training and practice. In Byzantium and the Islamic world, many of the Greek works were translated into Arabic and many of the works of Aristotle were preserved.

De arte venandi, by Frederick II, Holy Roman Emperor, was an influential medieval natural history text that explored bird morphology.

Medieval Muslim physicians, scientists and philosophers made significant contributions to zoological knowledge between the 8th and 13th centuries during what is known as the "Islamic Golden Age" or "Muslim Agricultural Revolution". The Afro-Arab scholar al-Jahiz (781–869) described early evolutionary ideas such as the struggle for existence. He also introduced the idea of a food chain, and was an early adherent of environmental determinism.

During the High Middle Ages, a few European scholars such as Hildegard of Bingen, Albertus Magnus and Frederick II expanded the natural history canon. Magnus' *De animalibus libri XXVI* is not the only volume of his commentaries on natural history, but it was one of the most extensive studies of zoological observation published before modern times. The rise of European universities, though important for the development of physics and philosophy, had little impact on zoological scholarship.

Zoology as a Science

Renaissance and Early Modern Developments

Prior to the Renaissance, accounts of animals were often apocryphal and creatures were often described as "legendary." This period was succeeded by the age of collectors and travellers, when many of the stories were actually demonstrated as true when the living or preserved specimens were brought to Europe. Verification by collecting of things, instead of the accumulation of anecdotes, then became more common, and scholars developed a new faculty of careful observation. The Renaissance brought expanded interest in both empirical natural history and physiology. In 1543, Andreas Vesalius inaugurated the modern era of Western medicine with his seminal human anatomy treatise *De humani corporis fabrica*, which was based on dissection of corpses. Vesalius was the first in a series of anatomists who gradually replaced scholasticism with empiricism in physiology and medicine, relying on first-hand experience rather than authority and abstract reasoning. Bestiaries—a genre that combines both the natural and figurative knowledge of animals—also became more sophisticated. Conrad Gessner great zoological work, *Historiae animalium*, appeared in four volumes, 1551-1558, at Zürich, a fifth being issued in 1587. His works were the starting-point of modern zoology. Other major works were produced by William Turner, Pierre Belon, Guillaume Rondelet, and Ulisse Aldrovandi. Artists such as Albrecht Dürer and Leonardo da Vinci, often working with naturalists, were also interested in the bodies of animals and humans, studying physiology in detail and contributing to the growth of anatomical knowledge.

In the 17th century, the enthusiasts of the new sciences, the investigators of nature by means of observation and experiment, banded themselves into academies or societies for mutual support and discourse. The first founded of surviving European academies, the Academia Naturae Curiosorum (1651) especially confined itself to the description and illustration of the structure of plants and animals; eleven years later (1662) the Royal Society of London was incorporated by royal charter, having existed without a name or fixed organisation for seventeen years previously (from 1645). A little later the Academy of Sciences of Paris was established by Louis XIV, later still the Royal Society of Sciences in Uppsala was founded. Systematizing, naming and classifying dominated zoology throughout much of the 17th and 18th centuries. Carl Linnaeus published a basic taxonomy for the natural world in 1735 (variations of which have been in use ever since), and in the 1750s introduced scientific names for all his species. While Linnaeus conceived of species as unchanging parts of a designed hierarchy, the other great naturalist of the 18th century, Georges-Louis Leclerc, Comte de Buffon, treated species as artificial categories and living forms as malleable—even suggesting the possibility of common descent. Though he was writing in an era before evolution existed, Buffon is a key figure in the history of evolutionary thought; his "transformist" theory would influence the evolutionary theories of both Jean-Baptiste Lamarck and Charles Darwin.

Before the Age of Exploration, naturalists had little idea of the sheer scale of biological diversity. The discovery and description of new species and the collection of specimens became a passion of scientific gentlemen and a lucrative enterprise for entrepreneurs; many naturalists traveled the globe in search of scientific knowledge and adventure.

Extending the work of Vesalius into experiments on still living bodies (of both humans and animals), William Harvey and other natural philosophers investigated the roles of blood, veins and arteries. Harvey's *De motu cordis* in 1628 was the beginning of the end for Galenic theory, and alongside Santorio Santorio's studies of metabolism, it served as an influential model of quantitative approaches to physiology.

Impact of the Microscope

In the early 17th century, the micro-world of zoology was just beginning to open up. A few lens-makers and natural philosophers had been creating crude microscopes since the late 16th century, and Robert Hooke published the seminal *Micrographia* based on observations with his own compound microscope in 1665. But it was not until Antony van Leeuwenhoek's dramatic improvements in lensmaking beginning in the 1670s—ultimately producing up to 200-fold magnification with a single lens—that scholars discovered spermatozoa, bacteria, infusoria and the sheer strangeness and diversity of microscopic life. Similar investigations by Jan Swammerdam led to new interest in entomology and built the basic techniques of microscopic dissection and staining.

Debate over the flood described in the Bible catalyzed the development of paleontology; in 1669 Nicholas Steno published an essay on how the remains of living organisms could be trapped in layers of sediment and mineralized to produce fossils. Although Steno's ideas about fossilization were well known and much debated among natural philosophers, an organic origin for all fossils would not be accepted by all naturalists until the end of the 18th century due to philosophical and theological debate about issues such as the age of the earth and extinction.

18th century microscopes from the Musée des Arts et Métiers, Paris.

Advances in microscopy also had a profound impact on biological thinking. In the early 19th century, a number of biologists pointed to the central importance of the cell. In 1838 and 1839, Schleiden and Schwann began promoting the ideas that (1) the basic unit of organisms is the cell and (2) that individual cells have all the characteristics of life, though they opposed the idea that (3) all cells come from the division of other cells. Thanks to the work of Robert Remak and Rudolf Virchow, however, by the 1860s most biologists accepted all three tenets of what came to be known as cell theory.

In advance of on the Origin of Species

Up through the 19th century, the scope of zoology was largely divided between physiology, which investigated questions of form and function, and natural history, which was concerned with the diversity of life and interactions among different forms of life and between life and non-life. By 1900, much of these domains overlapped, while natural history (and its counterpart natural philosophy) had largely given way to more specialized scientific disciplines—cytology, bacteriology, morphology, embryology, geography, and geology. Widespread travel by naturalists in the early-to-mid-19th century resulted in a wealth of new information about the diversity and distribution of living organisms. Of particular importance was the work of Alexander von Humboldt, which analyzed the relationship between organisms and their environment (i.e., the domain of natural history) using the quantitative approaches of natural philosophy (i.e., physics and chemistry). Humboldt's work laid the foundations of biogeography and inspired several generations of scientists.

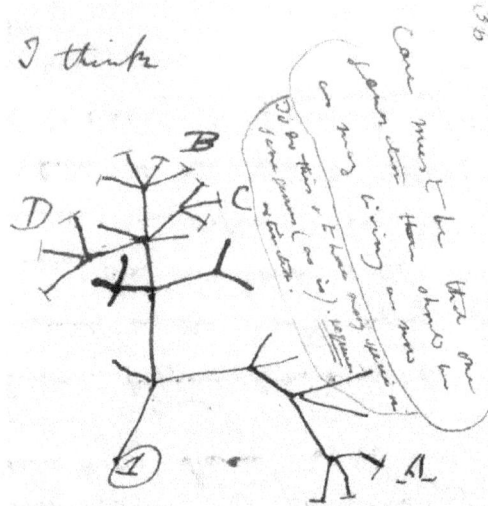

Charles Darwin's first sketch of an evolutionary tree from his *First Notebook on Transmutation of Species* (1837)

The emerging discipline of geology also brought natural history and natural philosophy closer together; Georges Cuvier and others made great strides in comparative anatomy and paleontology in the late 1790s and early 19th century. In a series of lectures and papers that made detailed comparisons between living mammals and fossil remains Cuvier was able to establish that the fossils were remains of species that had become extinct—rather than being remains of species still alive elsewhere in the world, as had been widely believed. Fossils discovered and described by Gideon Mantell, William Buckland, Mary Anning, and Richard Owen among others helped establish that there had been an 'age of reptiles' that had preceded even the prehistoric mammals. These discoveries captured the public imagination and focused attention on the history of life on earth.

The most significant evolutionary theory before Darwin's was that of Jean-Baptiste Lamarck; based on the inheritance of acquired characteristics (an inheritance mechanism that was widely accepted until the 20th century), it described a chain of development stretching from the lowliest microbe to humans. The British naturalist Charles Darwin, combining the biogeographical approach of Humboldt, the uniformitarian geology of Lyell, Thomas Malthus's writings on population growth, and his own morphological expertise, created a more successful evolutionary theory based on natural selection; similar evidence led Alfred Russel Wallace to independently reach the same conclusions. Charles Darwin's early interest in nature led him on a five-year voyage on HMS *Beagle* which established him as an eminent geologist whose observations and theories supported Charles Lyell's uniformitarian ideas, and publication of his journal of the voyage made him famous as a popular author. Puzzled by the geographical distribution of wildlife and fossils he collected on the voyage, Darwin investigated the transmutation of species and conceived his theory of natural selection in 1838. Although he discussed his ideas with several naturalists, he needed time for extensive research and his geological work had priority. He was writing up his theory in 1858 when Alfred Russel Wallace sent him an essay which described the same idea, prompting immediate joint publication of both of their theories. Darwin's *On the Origin of Species*, published on 24 November 1859, a seminal work of scientific literature, was to be the foundation of evolutionary biology.

History of Zoology Since 1859

This article considers the history of zoology since the theory of evolution by natural selection proposed by Charles Darwin in 1859.

Charles Darwin gave new direction to morphology and physiology, by uniting them in a common biological theory: the theory of organic evolution. The result was a reconstruction of the classification of animals upon a genealogical basis, fresh investigation of the development of animals, and early attempts to determine their genetic relationships. The end of the 19th century saw the fall of spontaneous generation and the rise of the germ theory of disease, though the mechanism of inheritance remained a mystery. In the early 20th century, the rediscovery of Mendel's work led to the rapid development of genetics by Thomas Hunt Morgan and his students, and by the 1930s the combination of population genetics and natural selection in the "neo-Darwinian synthesis".

Second Half of Nineteenth Century

Darwin and the Theory of Evolution

The 1859 publication of Darwin's theory in *On the Origin of Species by Means of Natural Selection, or the Preservation of Favoured Races in the Struggle for Life* is often considered the central event in the history of modern zoology. Darwin's established credibility as a naturalist, the sober tone of the work, and most of all the sheer strength and volume of evidence presented, allowed *Origin* to succeed where previous evolutionary works such as the anonymous *Vestiges of Creation* had failed. Most scientists were convinced of evolution and common descent by the end of the 19th century. However, natural selection would not be accepted as the primary mechanism of evolution until well into the 20th century, as most contemporary theories of heredity seemed incompatible with the inheritance of random variation.

Alfred Russel Wallace, following on earlier work by de Candolle, Humboldt and Darwin, made major contributions to zoogeography. Because of his interest in the transmutation hypothesis, he paid particular attention to the geographical distribution of closely allied species during his field work first in South America and then in the Malay archipelago. While in the archipelago he identified the Wallace line, which runs through the Spice Islands dividing the fauna of the archipelago between an Asian zone and a New Guinea/Australian zone. His key question, as to why the fauna of islands with such similar climates should be so different, could only be answered by considering their origin. In 1876 he wrote *The Geographical Distribution of Animals*, which was the standard reference work for over half a century, and a sequel, *Island Life*, in 1880 that focused on island biogeography. He extended the six-zone system developed by Philip Sclater for describing the geographical distribution of birds to animals of all kinds. His method of tabulating data on animal groups in geographic zones highlighted the discontinuities; and his appreciation of evolution allowed him to propose rational explanations, which had not been done before.

The scientific study of heredity grew rapidly in the wake of Darwin's *Origin of Species* with the work of Francis Galton and the biometricians. The origin of genetics is usually traced to the 1866 work of the monk Gregor Mendel, who would later be credited with the laws of inheritance. However, his work was not recognized as significant until 35 years afterward. In the meantime, a variety of theories of inheritance (based on pangenesis, orthogenesis, or other mechanisms) were debated and investigated vigorously.

In 1859, Charles Darwin placed the whole theory of organic evolution on a new footing, by his discovery of a process by which organic evolution can occur, and provided observational evidence that it had done so. This changed the attitudes of most exponents of the scientific method. Darwin's discoveries revolutionised the zoological and botanical sciences, by introducing the theory of evolution by natural selection as an explanation for the diversity of all animal and plant life. The subject-matter of this new science, or branch of biological science, had been neglected: it did not form part of the studies of the collector and systematist, nor was it a branch of anatomy, nor of the physiology pursued by medical men, nor again was it included in the field of microscopy and the cell theory. Almost a thousand years before Darwin, the Arab scholar Al-Jahiz (781 – 868) had already developed a rudimentary theory of natural selection , describing the Struggle for existence in his *Book of Animals* where he speculates on how environmental factors can affect the characteristics of species by forcing them to adapt and then passing on those new traits to future generations. However, his work had largely been forgotten, along with many other early advances of Arab scientists, and there is no evidence that his works were known to Darwin.

The area of biological knowledge which Darwin was the first to subject to scientific method and to render, as it were, contributory to the great stream formed by the union of the various branches, is that which relates to the breeding of animals and plants, their congenital variations, and the transmission and perpetuation of those variations. This branch of biological science may be called thremmatology - the science of breeding. Outside the scientific world, an immense mass of observation and experiment had grown up in relation to this subject. From the earliest times the shepherd, the farmer, the horticulturist, and the fancier had for practical purposes made themselves acquainted with a number of biological laws, and successfully applied them without exciting more than an occasional notice from the academic students of biology. Darwin made use of these observations and formulated their results to a large extent as the laws of variation and heredity. As the breeder selects a congenital variation which suits his requirements, and by breeding from the animals (or plants) exhibiting that variation obtains a new breed specially characterised by that variation, so in nature is there a selection amongst all the congenital variations of each generation of a species. This selection depends on the fact that more young are born than the natural provision of food will support. In consequence of this excess of births there is a struggle for existence and a survival of the fittest, and consequently an ever-present necessarily acting selection, which either maintains accurately the form of the species from generation to generation or leads to its modification in correspondence with changes in the surrounding circumstances which have relation to its fitness for success in the struggle for life, structures to the service of the organisms in which they occur.

It cannot be said that previously to Darwin there had been any very profound study of teleology, but it had been the delight of a certain type of mind, that of the lovers of nature or naturalists par excellence as they were sometimes termed, to watch the habits of living animals and plants and to point out the remarkable ways in which the structure of each variety of organic life was adapted to the special circumstances of life of the variety or species. The astonishing colours and grotesque forms of some animals and plants which the museum zoologists gravely described without comment were shown by these observers of living nature to have their significance in the economy of the organism possessing them; and a general doctrine was recognized, to the effect that no part or structure of an organism is without definite use and adaptation, being designed by the Creator for the benefit of the creature to which it belongs, or else for the benefit, amusement or instruction of his highest creatureman. Teleology in this form of the doctrine of design was never very deeply rooted amongst scientific anatomists and systematists. It was considered permissible to speculate somewhat vaguely on the subject of the utility of this or that startling variety of structure; but few attempts, though some of great importance, were made systematically to explain by observation and experiment the adaptation of organic structures to particular purposes in the case of the lower animals and plants. Teleology had, indeed, an important part in the development of physiology - the knowledge of the mechanism, the physical and chemical properties, of the parts of the body of man and the higher animals allied to him. But, as applied to lower and more obscure forms of life, teleology presented almost insurmountable difficulties; and consequently, in place of exact experiment and demonstration, the most reckless though ingenious assumptions were made as to the utility of the parts and organs of lower animals.

Darwin's theory had as one of its results the reformation and rehabilitation of teleology. According to that theory, every organ, every part, colour and peculiarity of an organism, must either be of benefit to that organism itself or have been so to its ancestors: no peculiarity of structure or general conformation, no habit or instinct in any organism, can be supposed to exist for the benefit or amusement of another organism, not even for the delectation of man himself.

A very subtle and important qualification of this generalization has to be recognized (and was recognized by Darwin) in the fact that owing to the interdependence of the parts of the bodies of living things and their profound chemical interactions and peculiar structural balance (what is called organic polarity) the variation of one single part (a spot of colour, a tooth, a claw, a leaflet) may, and demonstrably does in many cases entail variation of other parts what are called correlated variations. Hence many structures which are obvious to the eye, and serve as distinguishing marks of separate species, are really not themselves of value or use, but are the necessary concomitants of less obvious and even altogether obscure qualities, which are the real characters upon which selection is acting. Such correlated variations may attain to great size and complexity without being of use. But eventually they may in turn become, in changed conditions, of selective value. Thus in many cases the difficulty of supposing that selection has acted on minute and imperceptible initial variations, so small as to have no selective value, may be got rid of. A useless correlated variation may have attained great volume and quality before it is (as it were) seized upon and perfected by natural selection. All organisms are essentially and necessarily built up by such correlated variations.

Necessarily, according to the theory of natural selection, structures either are present because they are selected as useful or because they are still inherited from ancestors to whom they were useful, though no longer useful to the existing representatives of those ancestors. Structures previously inexplicable were now explained as survivals from a past age, no longer useful though once of value. Every variety of form and colour was urgently and absolutely called upon to produce its title to existence either as an active useful agent or as a survival. Darwin himself spent a large part of the later years of his life in thus extending the new teleology.

The old doctrine of types, which was used by the philosophically minded zoologists (and botanists) of the first half of the 19th century as a ready means of explaining the failures and difficulties of the doctrine of design, fell into its proper place under the new dispensation. The adherence to type, the favourite conception. of the transcendental morphologist, was seen to be nothing more than the expression of one of the laws of thremmatology, the persistence of hereditary transmission of ancestral characters, even when they have ceased to be significant or valuable in the struggle for existence, whilst the so-called evidences of design which was supposed to modify the limitations of types assigned to Himself by the Creator were seen to be adaptations due to the selection and intensification by selective breeding of fortuitous congenital variations, which happened to prove more useful than the many thousand other variations which did not survive in the struggle for existence.

Thus not only did Darwins theory give a new basis to the study of organic structure, but, whilst rendering the general theory of organic evolution equally acceptable and necessary, it explained the existence of low and simple forms of life as survivals of the earliest ancestry of the more highly complex forms, and revealed the classifications of the systematist as unconscious attempts to construct the genealogical tree or pedigree of plants and animals. Finally, it brought the simplest living matter or formless protoplasm before the mental vision as the starting point whence, by the operation of necessary mechanical causes, the highest forms have been evolved, and it rendered unavoidable the conclusion that this earliest living material was itself evolved by gradual processes, the result also of the known and recognized laws of physics and chemistry, from material which we should call not living. It abolished the conception of life as an entity above and beyond the common properties of matter, and led to the conviction that the marvellous and exceptional qualities of that which we call living matter are nothing more nor less than an exceptionally complicated development of those chemical and physical properties which we recognize in a gradually ascending scale of evolution in the carbon compounds, containing nitrogen as well as oxygen, sulphur and hydrogen as constituent atoms of their enormous molecules. Thus mysticism was finally banished from the domain of biology, and zoology became one of the physical sciencesthe science which seeks to arrange and discuss the phenomena of animal life and form, as the outcome of the operation of the laws of physics and chemistry.

A subdivision of zoology which was at one time in favour is simply into morphology and physiology, the study of form and structure on the one hand, and the study of the activities and functions of the forms and structures of the other. But a logical division like this is not necessarily conducive to the ascertainment and remembrance of the historical progress and present significance of the science. No such distinction of mental activities as that involved in the division of the study of animal life into morphology and physiology has ever really existed: the investigator of animal forms has never entirely ignored the functions of the forms studied by him, and the experimental inquirer into the functions and properties of animal tissues and organs has always taken very careful account of the forms of those tissues and organs. A more instructive subdivision must be one which corresponds to the separate currents of thought and mental preoccupation which have been historically manifested in western Europe in the gradual evolution of what is to-day the great river of zoological doctrine to which they have all been rendered contributory.

Cell theory, Embryology and Germ Theory

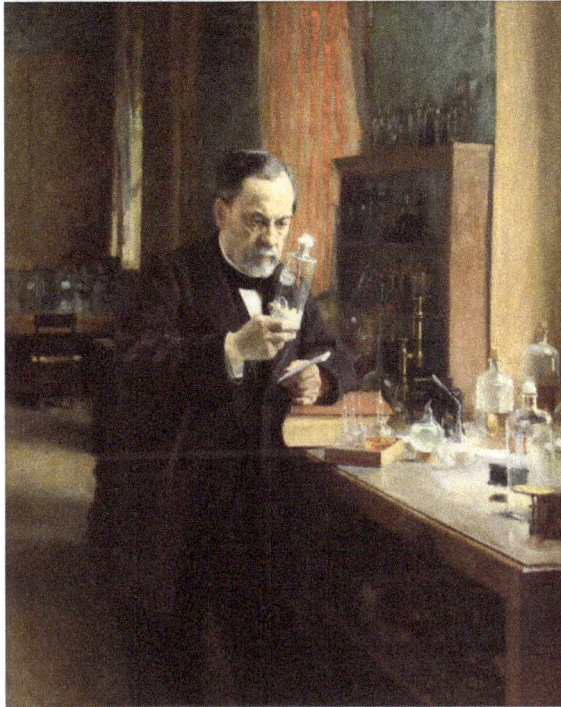

Innovative laboratory glassware and experimental methods developed by Louis Pasteur and other biologists contributed to the young field of bacteriology in the late 19th century.

Cell theory led zoologists to re-envision individual organisms as interdependent assemblages of individual cells. Scientists in the rising field of cytology, armed with increasingly powerful microscopes and new staining methods, soon found that even single cells were far more complex than the homogeneous fluid-filled chambers described by earlier microscopists. Much of the research on cell reproduction came together in August Weismann's theory of heredity: he identified the nucleus (in particular chromosomes) as the hereditary material, proposed the distinction between somatic cells and germ cells (arguing that chromosome number must be halved for germ cells, a precursor to the concept of meiosis), and adopted Hugo de Vries's theory of pangenes. Weismannism was extremely influential, especially in the new field of experimental embryology.

By the 1880s, bacteriology was becoming a coherent discipline, especially through the work of Robert Koch, who introduced methods for growing pure cultures on agar gels containing specific nutrients in Petri dishes. The long-held idea that living organisms could easily originate from nonliving matter (spontaneous generation) was attacked in a series of experiments carried out by Louis Pasteur, while debates over vitalism vs. mechanism (a perennial issue since the time of Aristotle and the Greek atomists) continued apace.

Physiology

Over the course of the 19th century, the scope of physiology expanded greatly, from a primarily medically oriented field to a wide-ranging investigation of the physical and chemical processes of life—including plants, animals, and even microorganisms in addition to man. *Living things as machines* became a dominant metaphor in biological (and social) thinking. Physiologists such as

Claude Bernard explored (through vivisection and other experimental methods) the chemical and physical functions of living bodies to an unprecedented degree, laying the groundwork for endocrinology (a field that developed quickly after the discovery of the first hormone, secretin, in 1902), biomechanics, and the study of nutrition and digestion. The importance and diversity of experimental physiology methods, within both medicine and zoology, grew dramatically over the second half of the 19th century. The control and manipulation of life processes became a central concern, and experiment was placed at the center of biological education.

Twentieth Century

At the beginning of the 20th century, zoological research was largely a professional endeavour. Most work was still done in the natural history mode, which emphasized morphological and phylogenetic analysis over experiment-based causal explanations. However, anti-vitalist experimental physiologists and embryologists, especially in Europe, were increasingly influential. The tremendous success of experimental approaches to development, heredity, and metabolism in the 1900s and 1910s demonstrated the power of experimentation in biology. In the following decades, experimental work replaced natural history as the dominant mode of research.

Early 20th Century Work (Variation and Heredity)

After publication of his work *The Origin of Species*, Darwin became interested in the animal and plant mechanisms that confer advantages to individual members of a species. Much important work was done by Fritz Muller (*Für Darwin*), by Hermann Müller (*Fertilization of Plants by Insects*), August Weismann, Edward B. Poulton and Abbott Thayer. There was considerable progress during this period in the field that would become known as genetics, the laws of variation and heredity (originally known as *thremmatology*). The progress of microscopy gave a clearer understanding of the origin of the egg-cell and sperm-cell and the process of fertilization.

Mendel and Zoology

Mendel's experiments on cultivated varieties of plants were published in 1865, but attracted little notice until thirty-five years later, sixteen years after his death. Mendel tried to gain a better understanding of heredity. His main experiments were with varieties of the edible pea. He chose a variety with one marked structural feature and crossed it with another variety in which that feature was absent. For example, he hybridized a tall variety, with a dwarf variety, a yellow-seeded variety with a green-seeded variety, and a smooth-seeded variety with a wrinkle-seeded variety. In each experiment, he concentrated on one character; after obtaining a first hybrid generation, he allowed the hybrids to self-fertilize, and recorded the number of individuals in the first, second, third, and fourth generations in which the chosen character appeared.

In the first hybrid generation, nearly all the individuals had the positive character, but in subsequent generations the positive character was not present in all individuals: half had the character and half did not. Thus the random pairing of two groups of reproductive cells yielded the proportion 1 PP, 2 PN, 1 NN, where P stands for the character and N for its absence - the character was present in three-quarters of the offspring and absent from a quarter. The failure of the character to distribute itself among all of the reproductive cells of a hybrid individual, and the limitation of its distribution to half only of those cells, prevents the swamping of a new character by interbreeding.

The tendency of the proportions in the offspring is to give, in a series of generations, a reversion from the hybrid form PN to a race with the positive character and a race without it. This tendency favours the persistence of a new character of large volume suddenly appearing in a stock. The observations of Mendel thus favoured the view that the variations upon which natural selection acts are not small but large and discontinuous. However, it did not appear that large variations would be favoured any more than small ones, or that the eliminating action of natural selection upon an unfavourable variation could be checked.

Much confusion arose in discussions of this topic, because of defective nomenclature. Some authors used the word *mutation* only for large variations that appeared suddenly and that could be inherited, and *fluctuation* for small variations, whether they could be transmitted or not. Other authors used *fluctuation* only for small, acquired variations due to changes in food, moisture and other features of the environment. This kind of variation is not heritable, but the small variations Darwin thought important are. The best classification of the variations in organisms separates those that arise from congenital variations from those that arise from variations of the environment or the food-supply. The former are innate variations, the latter are "acquired variations". Both innate and acquired variations include some that are more and some that are less obvious. There are slight innate variations in every new generation of every species; their greatness or smallness so far as human perception goes is not of much significance, their importance for the origin of new species depends on whether they are valuable to the organism in the struggle for existence and reproduction. An imperceptible physiological difference might be of selective value, and it might carry with it correlated variations that may or may not appeal to the human eye, but are of no selective value themselves.

The views of Hugo de Vries and others about the importance of saltatory variation, the soundness of which was still not generally accepted in 1910, may be gathered from the article Mendelism. A due appreciation of the far-reaching results of correlated variation must, it appeared, give a new and distinct explanation of large mutations, discontinuous variation, and saltatory evolution. The analysis of the specific variations of organic form to determine the nature and limitation of a single character, and whether two variations of a structural unit can blend when one is transmitted by the male parent and the other by the female, were yet to be determined. It was not clear whether absolute blending was possible, or whether all apparent blending was only a more-or-less minutely subdivided mosaic of non-combinable characters of the parents.

Another important development of Darwin's conclusions deserves notice. The fact of variation was familiar: no two animals, even of the same brood, are alike. Jean-Baptiste Lamarck hypothesised that structural alterations acquired by a parent might be transmitted to the offspring, and as these are acquired by an animal or plant as a consequence of the action of the environment, the offspring would sometimes start with a greater fitness for those conditions than its parents started with. In turn, it would acquire a greater development of the same modification, which it would transmit to its offspring. Lamarck argued that, over several generations, a structural alteration might thus be acquired. The familiar illustration of Lamarck's hypothesis is that of the giraffe, whose long neck might, he suggested, was acquired by the efforts of a short-necked race of herbivores who stretched their necks to reach the foliage of trees in a land where grass was deficient, the effort producing a longer neck of each generation, which was then transmitted to the next. This process is known as 'direct adaptation'.

Such structural adaptations are acquired by an animal in the course of its life, but are limited in degree and rare, rather than frequent and obvious. Whether acquired characters could be transmitted to the next generation was a very different issue. Darwin excluded any assumption of the transmission of acquired characters. He pointed to the fact of congenital variation, and showed that congenital variations are arbitrary and non-significant.

Congenital Variation

At the beginning of the 20th century, the causes of congenital variation were obscure, although it was recognised that they were largely due to a mixing of the matter that constituted the fertilized germ or embryo-cell from two individuals. Darwin had shown that congenital variation was all-important. A popular illustration of the difference was this: a man born with four fingers only on his right hand might transmit this peculiarity to at least some of his children; but a man with one finger chopped off will produce children with five fingers. Darwin, influenced by some facts that seemed to favour the Lamarckian hypothesis, thought that acquired characters are sometimes transmitted, but did not consider that this mechanism was likely to be of great importance.

After Darwin's writings, there was an effort to find evidence for the transmission of acquired characters; ultimately, the Lamarckian hypothesis of transmission of acquired characters was not supported by evidence, and was dismissed. August Weismann argued from the structure of the egg-cell and sperm-cell, and from how and when they are derived in the growth of the embryo from the egg, that it was impossible that a change in parental structure could produce a representative change in the germ or sperm-cells.

The only evidence that seemed to support the Lamarckian hypothesis were the experiments of Charles Brown-Séquard, who produced epilepsy in guinea-pigs by bisection of the large nerves or spinal cord, which led him to believe that, in rare instances, the artificially-produced epilepsy and mutilation of the nerves was transmitted. The record of Brown-Séquard's original experiments was unsatisfactory, and attempts reproduce them were unsuccessful. Conversely, the vast number of experiments in the cropping of the tails and ears of domestic animals, as well as of similar operations on man, had negative results. Stories of tailess kittens, puppies, and calves, born from parents one of whom had been thus injured, are abundant, but failed to stand experimental examination.

Whilst evidence of the transmission of an acquired character proved wanting, the *a priori* arguments in its favour were recognized as flawed, and cases that appeared to favour the Lamarckian assumption were found to be better explained by the Darwinian principle. For example, the occurrence of blind animals in caves and in the deep sea was a fact that even Darwin regarded as best explained by the atrophy of the eye in successive generations through the absence of light and consequent disuse. However, it was suggested that this is better explained by natural selection acting on congenital fortuitous variations. Some animals are born with distorted or defective eyes. If a number of some species of fish are swept into a cavern, those with perfect eyes would follow the light and eventually escape, leaving behind those with imperfect eyes to breed in the dark place. In every succeeding generation this would be the case, and even those with weak but still seeing eyes would escape, until only a pure race of blind animals would be left in the cavern.

Transmission

It was argued that the elaborate structural adaptations of the nervous system that underlie instincts must have been slowly built up by the transmission to offspring of acquired experience. It seemed hard to understand how complicated instincts could be due to the selection of congenital variations, or be explained except by the transmission of habits acquired by the parent. However, imitation of the parent by the young account for some, and there are cases in which elaborate actions must be due to the natural selection of a fortuitously-developed habit. Such cases are the habits of 'shamming dead' and the combined posturing and colour peculiarities of certain caterpillars (Lepidoptera larvae) that cause them to resemble dead twigs or similar objects. The advantage to the caterpillar is that it escapes (say) a bird that would, were it not deceived, attack and eat it. Preceding generations of caterpillars cannot have acquired this habit of posturing by experience; either a caterpillar postures and escapes, or it does not posture and is eaten - it is not half eaten and allowed to profit by experience. Thus, we seem justified in assuming that there are many movements of stretching and posturing possible to caterpillars, that some had a fortuitous tendency to one position, some to another, and, that among all the variety of habitual movements, one is selected and perpetuated because it happened to make the caterpillar look more like a twig.

Record of the Past

Man, compared with other animals, has the fewest instincts and the largest brain in proportion to body size. He builds up, from birth onwards, his own mental mechanisms, and forms more of them, and takes longer in doing so, than any other animal. The later stages of evolution from ape-like ancestors have consisted in the acquisition of a larger brain and in the education of that brain. A new feature in organic development makes its appearance when we set out the facts of man's evolutionary history. This factor is the *record of the past*, which grows and develops by laws other than those affecting the perishable bodies of successive generations of mankind, so that man, by the interaction of the *record* and his educability, is subject to laws of development unlike those by which the rest of the living world is governed.

Ecology and Environmental Science

In the early 20th century, naturalists were faced with increasing pressure to add rigor and preferably experimentation to their methods, as the newly prominent laboratory-based biological disciplines had done. Ecology had emerged as a combination of biogeography with the biogeochemical cycle concept pioneered by chemists; field biologists developed quantitative methods such as the quadrat and adapted laboratory instruments and cameras for the field to further set their work apart from traditional natural history. Zoologists did what they could to mitigate the unpredictability of the living world, performing laboratory experiments and studying semi-controlled natural environments; new institutions like the Carnegie Station for Experimental Evolution and the Marine Biological Laboratory provided more controlled environments for studying organisms through their entire life cycles.

Charles Elton's studies of animal food chains was pioneering among the succession of quantitative methods that colonized the developing ecological specialties. Ecology became an independent discipline in the 1940s and 1950s after Eugene P. Odum synthesized many of the concepts of ecosystem ecology, placing relationships between groups of organisms (especially material and energy relationships) at the center of the field. In the 1960s, as evolutionary theorists explored the possibility of multiple units of selection, ecologists turned to evolutionary approaches. In population ecology, debate over group selection was brief but vigorous; by 1970, most zoologists agreed that natural selection was rarely effective above the level of individual organisms.

Classical Genetics, The Modern Synthesis, and Evolutionary Theory

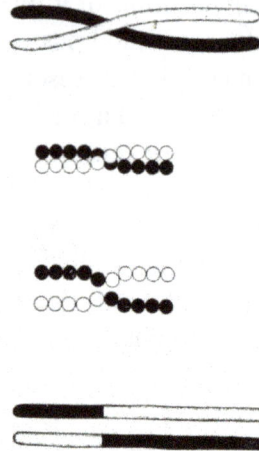

FIG. 64. Scheme to illustrate a method of crossing over of the chromosomes.

Thomas Hunt Morgan's illustration of crossing over, part of the Mendelian-chromosome theory of heredity

1900 marked the so-called *rediscovery of Mendel*: Hugo de Vries, Carl Correns, and Erich von Tschermak independently arrived at Mendel's laws (which were not actually present in Mendel's work). Soon after, cytologists (cell biologists) proposed that chromosomes were the hereditary material. Between 1910 and 1915, Thomas Hunt Morgan and the "Drosophilists" in his fly lab forged these two ideas—both controversial—into the "Mendelian-chromosome theory" of heredity. They quantified the phenomenon of genetic linkage and postulated that genes reside on chromosomes like beads on string; they hypothesized crossing over to explain linkage and constructed genetic maps of the fruit fly *Drosophila melanogaster*, which became a widely used model organism.

Hugo de Vries tried to link the new genetics with evolution; building on his work with heredity and hybridization, he proposed a theory of mutationism, which was widely accepted in the early 20th century. Lamarckism also had many adherents. Darwinism was seen as incompatible with the continuously variable traits studied by biometricians, which seemed only partially heritable. In the 1920s and 1930s—following the acceptance of the Mendelian-chromosome theory— the emergence of the discipline of population genetics, with the work of R.A. Fisher, J.B.S. Haldane and Sewall Wright, unified the idea of evolution by natural selection with Mendelian genetics, producing the modern synthesis. The inheritance of acquired characters was rejected, while mutationism gave way as genetic theories matured.

In the second half of the century the ideas of population genetics began to be applied in the new discipline of the genetics of behavior, sociobiology, and, especially in humans, evolutionary psychology. In the 1960s W.D. Hamilton and others developed game theory approaches to explain altruism from an evolutionary perspective through kin selection. The possible origin of higher organisms through endosymbiosis, and contrasting approaches to molecular evolution in the gene-centered view (which held selection as the predominant cause of evolution) and the neutral theory (which made genetic drift a key factor) spawned perennial debates over the proper balance of adaptationism and contingency in evolutionary theory.

In the 1970s Stephen Jay Gould and Niles Eldredge proposed the theory of punctuated equilibrium which holds that stasis is the most prominent feature of the fossil record, and that most evolutionary changes occur rapidly over relatively short periods of time. In 1980 Luis Alvarez and Walter Alvarez proposed the hypothesis that an impact event was responsible for the Cretaceous–Paleogene extinction event. Also in the early 1980s, statistical analysis of the fossil record of marine organisms published by Jack Sepkoski and David M. Raup lead to a better appreciation of the importance of mass extinction events to the history of life on earth.

Twenty-first Century

Advances were made in analytical chemistry and physics instrumentation including improved sensors, optics, tracers, instrumentation, signal processing, networks, robots, satellites, and compute power for data collection, storage, analysis, modeling, visualization, and simulations. These technology advances allowed theoretical and experimental research including internet publication of zoological science. This enabled worldwide access to better measurements, theoretical models, complex simulations, theory predictive model experimentation, analysis, worldwide internet observational data reporting, open peer-review, collaboration, and internet publication.

References

- Needham, Joseph; Ronan, Colin Alistair (1995). The Shorter Science and Civilisation in China: An Abridgement of Joseph Needham's Original Text, Vol. 1. Cambridge University Press. p. 101. ISBN 0-521-29286-7.

- Paul S. Agutter & Denys N. Wheatley (2008). Thinking about Life: The History and Philosophy of Biology and Other Sciences. Springer. p. 43. ISBN 1-4020-8865-5.

Sub-disciplines of Zoology

The progress that the natural sciences has made in the last hundred and fifty years has no parallel. Such a radical change in the sciences has created new ways of inference and deduction. The chapter strategically encompasses and incorporates the major components and key concepts of zoology, providing a complete understanding.

Physiology

Physiology is the scientific study of the normal function in living systems. A sub-discipline of biology, its focus is in how organisms, organ systems, organs, cells, and biomolecules carry out the chemical or physical functions that exist in a living system. Given the size of the field, it is divided into, among others, animal physiology (including that of humans), plant physiology, cellular physiology, microbial physiology (microbial metabolism), bacterial physiology, and viral physiology. The Nobel Prize in Physiology or Medicine is awarded to those who make significant achievements in this discipline by the Royal Swedish Academy of Sciences. In medicine, a physiologic state is one occurring from normal body function, rather than pathologically, which is centered on the abnormalities that occur in animal diseases, including humans.

Oil painting depicting Claude Bernard, the father of modern physiology, with his pupils

History

Physiological studies date back to ancient civilizations of India and Egypt alongside anatomical studies, but did not utilize dissection or vivisection.

The study of human physiology as a medical field dates back to at least 420 BC to the time of Hippocrates, also known as the "father of medicine." Hippocrates incorporated his belief system called the theory of humours, which consisted of four basic substance: earth, water, air and fire. Each substance is known for having a corresponding humour: black bile, phlegm, blood and yellow bile, respectively. Hippocrates also noted some emotional connections to the four humours, which Claudis Galenus would later expand on. The critical thinking of Aristotle and his emphasis on the relationship between structure and function marked the beginning of physiology in Ancient Greece. Like Hippocrates, Aristotle took to the humoral theory of disease, which also consisted of four primary qualities in life: hot, cold, wet and dry. Claudius Galenus (c. ~130–200 AD), known as Galen of Pergamum, was the first to use experiments to probe the functions of the body. Unlike Hippocrates though, Galen argued that humoral imbalances can be located in specific organs, including the entire body. His modification of this theory better equipped doctors to make more precise diagnoses. Galen also played off of Hippocrates idea that emotions were also tied to the humours, and added the notion of temperaments: sanguine corresponds with blood; phlegmatic is tied to phlegm; yellow bile is connected to choleric; and black bile corresponds with melancholy. Galen also saw the human body consisting of three connected systems: the brain and nerves, which are responsible for thoughts and sensations; the heart and arteries, which give life; and the liver and veins, which can be attributed to nutrition and growth. Galen was also the founder of experimental physiology. And for the next 1,400 years, Galenic physiology was a powerful and influential tool in medicine.

Jean Fernel (1497–1558), a French physician, introduced the term "physiology".

In 1858, Joseph Lister studied the cause of blood coagulation and inflammation that resulted after previous injuries and surgical wounds. He later discovered and implemented antiseptics in the operating room, and as a result decreases death rate from surgery by a substantial amount.

In 1891, Ivan Pavlov performed research on "conditional reflexes" that involved dogs' saliva production in response to a plethora of sounds and visual stimuli.

In the 19th century, physiological knowledge began to accumulate at a rapid rate, in particular with the 1838 appearance of the Cell theory of Matthias Schleiden and Theodor Schwann. It radically stated that organisms are made up of units called cells. Claude Bernard's (1813–1878) further discoveries ultimately led to his concept of *milieu interieur* (internal environment), which would later be taken up and championed as "homeostasis" by American physiologist Walter B. Cannon in 1929. By homeostasis, Cannon meant "the maintenance of steady states in the body and the physiological processes through which they are regulated." In other words, the body's ability to regulate its internal environment. It should be noted that, William Beaumont was the first American to utilize the practical application of physiology.

The Physiological Society was founded in London in 1876 as a dining club The American Physiological Society (APS) is a nonprofit devoted to fostering education, scientific research, and dissemination of information in the physiological sciences. The Society was founded in 1887 with 28 members.

In the 20th century, biologists became interested in how organisms other than human beings function, eventually spawning the fields of comparative physiology and ecophysiology. Major figures in these fields include Knut Schmidt-Nielsen and George Bartholomew. Most recently, evolutionary physiology has become a distinct subdiscipline.

In 1920, August Krogh won the Nobel Prize for discovering how, in capillaries, blood flow is regulated.

In 1954, Andrew Huxley and Hugh Huxley, alongside their research team, discovered the sliding filaments in skeletal muscle, known today as the sliding filament theory.

Women in Physiology

Initially, women were largely excluded from official involvement in any physiological society. The American Physiological Society, for example, was founded in 1887 and included only men in its ranks. In 1902, the American Physiological Society elected Ida Hyde as the first female member of the society. Hyde, a representative of the American Association of University Women and a global advocate for gender equality in education, attempted to promote gender equality in every aspect of science and medicine.

Soon thereafter, in 1913, J.S. Haldane proposed that women be allowed to formally join The Physiological Society, which had been founded in 1876. On 3 July 1915, six women were officially admitted: Florence Buchanan, Winifred Cullis, Ruth C. Skelton, Sarah C. M. Sowton, Constance Leetham Terry, and Enid M. Tribe. The centenary of the election of women was celebrated in 2015 with the publication of a book "Women physiologists: centenary celebrations and beyond for The Physiological Society ISBN 978-0-9933410-0-7.

Prominent women physiologists include:

- Gerty Cori, along with husband Carl Cori, received the Nobel Prize in Physiology or Medicine in 1947 for their discovery of the phosphate-containing form of glucose known as glycogen, as well as its function within eukaryotic metabolic mechanisms for energy production. Moreover, they discovered the Cori cycle, also known as the Lactic acid cycle, which describes how muscle tissue converts glycogen into lactic acid via lactic acid fermentation.

- Gertrude Elion, along with George Hitchings and Sir James Black, received the Nobel Prize for Physiology or Medicine in 1988 for their development of drugs employed in the treatment of several major diseases, such as leukemia, some autoimmune disorders, gout, malaria, and viral herpes.

- Linda B. Buck, along with Richard Axel, received the Nobel Prize in Physiology or Medicine in 2004 for their discovery of odorant receptors and the complex organization of the olfactory system.

- Françoise Barré-Sinoussi, along with Luc Montaginer, received the Nobel Prize in Physiology or Medicine in 2008 for their work on the identification of the Human Immunodeficiency Virus (HIV), the cause of Acquired Immunodeficiency Syndrome (AIDS).

- Elizabeth Blackburn, along with Carol Greider and Jack Szostak, was awarded the 2009

Nobel Prize for Physiology or Medicine for the discovery of the genetic composition and function of telomeres and the enzyme called telomerase.

Subdisciplines

There are many ways to categorize the subdiscplines of physiology:

- based on the taxa studied: human physiology, animal physiology, plant physiology, microbial physiology, viral physiology

- based on the level of organization: cell physiology, molecular physiology, systems physiology, organismal physiology, ecological physiology, integrative physiology

- based on the process that causes physiological variation: developmental physiology, environmental physiology, evolutionary physiology

- based on the ultimate goals of the research: applied physiology (e.g., medical physiology), non-applied (e.g., comparative physiology)

Human Physiology

Human physiology seeks to understand the mechanisms that work to keep the human body alive and functioning, through scientific enquiry into the nature of mechanical, physical, and biochemical functions of humans, their organs, and the cells of which they are composed. The principal level of focus of physiology is at the level of organs and systems within systems. The endocrine and nervous systems play major roles in the reception and transmission of signals that integrate function in animals. Homeostasis is a major aspect with regard to such interactions within plants as well as animals. The biological basis of the study of physiology, integration refers to the overlap of many functions of the systems of the human body, as well as its accompanied form. It is achieved through communication that occurs in a variety of ways, both electrical and chemical.

Much of the foundation of knowledge in human physiology was provided by animal experimentation. Physiology is the study of function and is closely related to anatomy which is the study of form and structure. Due to the frequent connection between form and function, physiology and anatomy are intrinsically linked and are studied in tandem as part of a medical curriculum.

Ethology

Ethology is the scientific and objective study of animal behaviour, usually with a focus on behaviour under natural conditions, and viewing behaviour as an evolutionarily adaptive trait. Behaviourism is a term that also describes the scientific and objective study of animal behaviour, usually referring to measured responses to stimuli or trained behavioural responses in a laboratory context, without a particular emphasis on evolutionary adaptivity. Many naturalists have studied aspects of animal behaviour throughout history. Ethology has its scientific roots in the work of Charles Darwin and of American and German ornithologists of the late 19th and early 20th century, including Charles O. Whitman, Oskar Heinroth, and Wallace Craig. The modern discipline of ethology is generally considered to have begun during the 1930s with the work of Dutch biologist Nikolaas Tinbergen and by Austrian biologists Konrad Lorenz and Karl von Frisch, joint awardees of the 1973 Nobel Prize in Physiology or Medicine. Ethology is a combination of laboratory and field science, with a strong relation to some other disciplines such as neuroanatomy, ecology, and evolutionary biology. Ethologists are typically interested in a behavioural process rather than in a particular animal group, and often study one type of behaviour, such as aggression, in a number of unrelated animals.

A range of animal behaviours

Ethology is a rapidly growing field. Since the dawn of the 21st century, many aspects of animal communication, emotions, culture, learning and sexuality that the scientific community long thought it understood have been re-examined, and new conclusions reached. New fields, such as neuroethology, have developed.

Understanding ethology or animal behaviour can be important in animal training. Considering the natural behaviours of different species or breeds enables the trainer to select the individuals best suited to perform the required task. It also enables the trainer to encourage the performance of naturally occurring behaviours and also the discontinuance of undesirable behaviours.

Etymology

The term *ethology* derives from the Greek language *ethos* meaning "character" and *-logia* meaning "the study of". The term was first popularized by American myrmecologist (a person who studies ants) William Morton Wheeler in 1902. An earlier, slightly different sense of the term was proposed by John Stuart Mill in his 1843 *System of Logic*. He recommended the development of a new science, "ethology", the purpose of which would be explanation of individual and national differences in character, on the basis of associationistic psychology. This use of the word was never adopted.

Relationship with Comparative Psychology

Comparative psychology also studies animal behaviour, but, as opposed to ethology, is construed as a sub-topic of psychology rather than as one of biology. Historically, where comparative psychology has included research on animal behaviour in the context of what is known about human psychology, ethology involves research on animal behaviour in the context of what is known about animal anatomy, physiology, neurobiology, and phylogenetic history. Furthermore, early comparative psychologists concentrated on the study of learning and tended to research behaviour in artificial situations, whereas early ethologists concentrated on behaviour in natural situations, tending to describe it as instinctive.

The two approaches are complementary rather than competitive, but they do result in different perspectives, and occasionally conflicts of opinion about matters of substance. In addition, for most of the twentieth century, comparative psychology developed most strongly in North America, while ethology was stronger in Europe. From a practical standpoint, early comparative psychologists concentrated on gaining extensive knowledge of the behaviour of very few species. Ethologists were more interested in understanding behaviour across a wide range of species to facilitate principled comparisons across taxonomic groups. Ethologists have made much more use of such cross-species comparisons than comparative psychologists have.

History

Scala Naturae and Lamarck's Theories

Jean-Baptiste Lamarck (1744–1829)

Until the 19th century, the most common theory among scientists was still the concept of *scala naturae*, proposed by Aristotle. According to this theory, living beings were classified on an ideal pyramid that represented non-living things (such as minerals and sediment) and the simplest animals on the lower levels, with complexity increasing progressively towards the top, occupied by human beings. In the Western world of the time, people believed animal species were eternal and immutable, created with a specific purpose, as this seemed the only possible explanation for the incredible variety of living beings and their surprising adaptation to their habitats.

Jean-Baptiste Lamarck (1744 - 1829) was the first biologist to describe a complex theory of evolution. His theory substantially comprised two statements: first, that animal organs and behaviour can change according to the way they are used; and second, that those characteristics can transmit from one generation to the next (the example of the giraffe whose neck becomes longer while trying to reach the upper leaves of a tree is well-known). The second statement is that every living organism, humans included, tends to reach a greater level of perfection. When Charles Darwin went to the Galapagos Islands, he was well aware of Lamarck's theories and was influenced by them.

Theory of Evolution by Natural Selection and the Beginnings of Ethology

Charles Darwin (1809–1882)

Because ethology is considered a topic of biology, ethologists have been concerned particularly with the evolution of behaviour and the understanding of behaviour in terms of the theory of natural selection. In one sense, the first modern ethologist was Charles Darwin, whose book *The Expression of the Emotions in Man and Animals* influenced many ethologists. He pursued his interest in behaviour by encouraging his protégé George Romanes, who investigated animal learning and intelligence using an anthropomorphic method, anecdotal cognitivism, that did not gain scientific support.

Other early ethologists, such as Charles O. Whitman, Oskar Heinroth, Wallace Craig and Julian Huxley, instead concentrated on behaviours that can be called instinctive, or natural, in that they occur in all members of a species under specified circumstances. Their beginning for studying the behaviour of a new species was to construct an ethogram (a description of the main types of behaviour with their frequencies of occurrence). This provided an objective, cumulative data-base of behaviour, which subsequent researchers could check and supplement.

Social Ethology and Recent Developments

In 1970, the English ethologist John H. Crook published an important paper in which he distinguished comparative ethology from social ethology, and argued that much of the ethology that had existed so far was really comparative ethology—examining animals as individuals—whereas, in the future, ethologists would need to concentrate on the behaviour of social groups of animals and the social structure within them.

Also in 1970, Robert Ardrey's book *The Social Contract: A Personal Inquiry into the Evolutionary Sources of Order and Disorder* was published. The book and study investigated animal behaviour and then compared human behaviour to it as a similar phenomenon.

E. O. Wilson's book *Sociobiology: The New Synthesis* appeared in 1975, and since that time, the study of behaviour has been much more concerned with social aspects. It has also been driven by the stronger, but more sophisticated, Darwinism associated with Wilson, Robert Trivers, and William Hamilton. The related development of behavioural ecology has also helped transform ethology. Furthermore, a substantial reapprochement with comparative psychology has occurred, so the modern scientific study of behaviour offers a more or less seamless spectrum of approaches: from animal cognition to more traditional comparative psychology, ethology, sociobiology, and behavioural ecology.

Growth of the Field

Due to the work of Lorenz and Tinbergen, ethology developed strongly in continental Europe during the years prior to World War II. After the war, Tinbergen moved to the University of Oxford, and ethology became stronger in the UK, with the additional influence of William Thorpe, Robert Hinde, and Patrick Bateson at the Sub-department of Animal Behaviour of the University of Cambridge, located in the village of Madingley. In this period, too, ethology began to develop strongly in North America.

Lorenz, Tinbergen, and von Frisch were jointly awarded the Nobel Prize in Physiology or Medicine in 1973 for their work of developing ethology.

Ethology is now a well-recognized scientific discipline, and has a number of journals covering developments in the subject, such as *Animal Behaviour, Animal Welfare, Applied Animal Behaviour Science, Behaviour, Behavioral Ecology* and *Journal of Ethology*. In 1972, the International Society for Human Ethology was founded to promote exchange of knowledge and opinions concerning human behaviour gained by applying ethological principles and methods and published their journal, *The Human Ethology Bulletin*. In 2008, in a paper published in the journal *Behaviour*, ethologist Peter Verbeek introduced the term "Peace Ethology" as a sub-discipline of Human Ethology that is concerned with issues of human conflict, conflict resolution, reconciliation, war, peacemaking, and peacekeeping behaviour.

Today, along with ethologists, many biologists, zoologists, primatologists, anthropologists, veterinarians, and physicians study ethology and other related fields such as animal psychology, the study of animal social groups, animal cognition and animal welfare science.

Instinct

Kelp gull chicks peck at red spot on mother's beak to stimulate regurgitating reflex

The Merriam-Webster dictionary defines instinct as "A largely inheritable and unalterable tendency of an organism to make a complex and specific response to environmental stimuli without involving reason".

Fixed Action Patterns

An important development, associated with the name of Konrad Lorenz though probably due more to his teacher, Oskar Heinroth, was the identification of fixed action patterns (FAPs). Lorenz popularized FAPs as instinctive responses that would occur reliably in the presence of identifiable stimuli called sign stimuli or "releasing stimuli". FAPs are now considered to be instinctive behavioural sequences that are relatively invariant within the species and almost inevitably run to completion.

One example of a releaser is the beak movements of many bird species performed by newly hatched chicks, which stimulates the mother to regurgitate food for her offspring. Other examples are the classic studies by Tinbergen on the egg-retrieval behaviour and the effects of a "supernormal stimulus" on the behaviour of graylag geese.

One investigation of this kind was the study of the waggle dance ("dance language") in bee communication by Karl von Frisch. Lorenz subsequently developed a theory of the evolution of animal communication based on his observations of fixed action patterns and the circumstances in which they are expressed.

Learning

Habituation

Habituation is a simple form of learning and occurs in many animal taxa. It is the process whereby an animal ceases responding to a stimulus. Often, the response is an innate behaviour. Essentially, the animal learns not to respond to irrelevant stimuli. For example, prairie dogs (*Cynomys ludovicianus*) give alarm calls when predators approach, causing all individuals in the group to quickly scramble down burrows. When prairie dog towns are located near trails used by humans, giving alarm calls every time a person walks by is expensive in terms of time and energy. Habituation to humans is therefore an important adaptation in this context.

Associative Learning

Associative learning in animal behaviour is any learning process in which a new response becomes associated with a particular stimulus. The first studies of associative learning were made by Russian physiologist Ivan Pavlov. Examples of associative learning include when a goldfish swims to the water surface when a human is going to feed it, or the excitement of a dog whenever it sees a leash as a prelude for a walk.

Imprinting

Example of imprinting in a moose

Being able to discriminate the members of one's own species is also of fundamental importance for reproductive success. Such discrimination can be based on a number of factors. However, this important type of learning only takes place in a very limited period of time. This kind of learning is called imprinting, and was a second important finding of Lorenz. Lorenz observed that the young of birds such as geese and chickens followed their mothers spontaneously from almost the first day after they were hatched, and he discovered that this response could be imitated by an arbitrary stimulus if the eggs were incubated artificially and the stimulus were presented during a critical period that continued for a few days after hatching.

Cultural Learning

Observational Learning

Imitation

Imitation is an advanced behaviour whereby an animal observes and exactly replicates the behaviour of another. The National Institutes of Health reported that capuchin monkeys preferred the company of researchers who imitated them to that of researchers who did not. The monkeys not only spent more time with their imitators but also preferred to engage in a simple task with them even when provided with the option of performing the same task with a non-imitator. Imitation has been observed in recent research on chimpanzees; not only did these chimps copy the actions of another individual, when given a choice, the chimps preferred to imitate the actions of the higher-ranking elder chimpanzee as opposed to the lower-ranking young chimpanzee.

Stimulus and Local Enhancement

There are various ways animals can learn using observational learning but without the process of imitation. One of these is *stimulus enhancement* in which individuals become interested in an object as the result of observing others interacting with the object. Increased interest in an object can result in object manipulation which allows for new object-related behaviours by trial-and-error learning. Haggerty (1909) devised an experiment in which a monkey climbed up the side of a cage, placed its arm into a wooden chute, and pulled a rope in the chute to release food. Another monkey was provided an opportunity to obtain the food after watching a monkey go through this process on four separate occasions. The monkey performed a different method and finally succeeded after trial-and-error. Another example familiar to some cat and dog owners is the ability of their animals to open doors. The action of humans operating the handle to open the door results in the animals becoming interested in the handle and then by trial-and-error, they learn to operate the handle and open the door.

In local enhancement, a demonstrator attracts an observer's attention to a particular location. Local enhancement has been observed to transmit foraging information among birds, rats and pigs. The stingless bee (*Trigona corvina*) uses local enhancement to locate other members of their colony and food resources.

Social Transmission

A well-documented example of social transmission of a behaviour occurred in a group of macaques on Hachijojima Island, Japan. The macaques lived in the inland forest until the 1960s, when a group of researchers started giving them potatoes on the beach: soon, they started venturing onto the beach, picking the potatoes from the sand, and cleaning and eating them. About one year later, an individual was observed bringing a potato to the sea, putting it into the water with one hand, and cleaning it with the other. This behaviour was soon expressed by the individuals living in contact with her; when they gave birth, this behaviour was also expressed by their young - a form of social transmission.

Teaching

Teaching is a highly specialized aspect of learning in which the "teacher" (demonstrator) adjusts their behaviour to increase the probability of the "pupil" (observer) achieving the desired end-result of the behaviour. For example, killer whales are known to intentionally beach themselves to catch pinniped prey. Mother killer whales teach their young to catch pinnipeds by pushing them onto the shore and encouraging them to attack the prey. Because the mother killer whale is altering her behaviour to help her offspring learn to catch prey, this is evidence of teaching. Teaching is not limited to mammals. Many insects, for example, have been observed demonstrating various forms of teaching to obtain food. Ants, for example, will guide each other to food sources through a process called "tandem running," in which an ant will guide a companion ant to a source of food.

It has been suggested that the pupil ant is able to learn this route to obtain food in the future or teach the route to other ants.This behaviour of teaching is also exemplified by crows. Specifically New Caledonian crows. The adults (whether individual or in families) teach their young adolescent offspring how to construct and utilize tools. For example; *Pandanus* branches are used to extract insects and other larvae from holes within trees.

Mating and the Fight for Supremacy

Individual reproduction is the most important phase in the proliferation of individuals or genes within a species: for this reason, there exist complex mating rituals, which can be very complex even if they are often regarded as FAPs. The stickleback's complex mating ritual, studied by Tinbergen, is regarded as a notable example of a FAP.

Often in social life, animals fight for the right to reproduce, as well as social supremacy. A common example of fighting for social and sexual supremacy is the so-called pecking order among poultry. Every time a group of poultry cohabitate for a certain time length, they establish a pecking order. In these groups, one chicken dominates the others and can peck without being pecked. A second chicken can peck all the others except the first, and so on. Higher level chickens are easily distinguished by their well-cured aspect, as opposed to lower level chickens. While the pecking order is establishing, frequent and violent fights can happen, but once established, it is broken only when other individuals enter the group, in which case the pecking order re-establishes from scratch.

Living in Groups

Several animal species, including humans, tend to live in groups. Group size is a major aspect of their social environment. Social life is probably a complex and effective survival strategy. It may be regarded as a sort of symbiosis among individuals of the same species: a society is composed of a group of individuals belonging to the same species living within well-defined rules on food management, role assignments and reciprocal dependence.

When biologists interested in evolution theory first started examining social behaviour, some apparently unanswerable questions arose, such as how the birth of sterile castes, like in bees, could be explained through an evolving mechanism that emphasizes the reproductive success of as many individuals as possible, or why, amongst animals living in small groups like squirrels, an individual would risk its own life to save the rest of the group. These behaviours may be examples of altruism. Of course, not all behaviours are altruistic, as indicated by the table below. For example, revengeful behaviour was at one point claimed to have been observed exclusively in *Homo sapiens*. However, other species have been reported to be vengeful, including reports of vengeful camels and chimpanzees.

Classification of social behaviours		
Type of behaviour	**Effect on the donor**	**Effect on the receiver**
Egoistic	Increases fitness	Decreases fitness
Cooperative	Increases fitness	Increases fitness
Altruistic	Decreases fitness	Increases fitness
Revengeful	Decreases fitness	Decreases fitness

Altruistic behaviour has been explained by the gene-centred view of evolution.

Benefits and Costs of Group Living

One advantage of group living can be decreased predation. If the number of predator attacks stays the same despite increasing prey group size, each prey may have a reduced risk of predator attacks through the dilution effect. Additionally, a predator that is confused by a mass of individuals can find it more difficult to single out one target. For this reason, the zebra's stripes offer not only camouflage in a habitat of tall grasses, but also the advantage of blending into a herd of other zebras. In groups, prey can also actively reduce their predation risk through more effective defense tactics, or through earlier detection of predators through increased vigilance.

Another advantage of group living can be an increased ability to forage for food. Group members may exchange information about food sources between one another, facilitating the process of resource location. Honeybees are a notable example of this, using the waggle dance to communicate the location of flowers to the rest of their hive. Predators also receive benefits from hunting in groups, through using better strategies and being able to take down larger prey.

Some disadvantages accompany living in groups. Living in close proximity to other animals can facilitate the transmission of parasites and disease, and groups that are too large may also experience greater competition for resources and mates.

Group Size

Theoretically, social animals should have optimal group sizes that maximize the benefits and minimize the costs of group living. However, in nature, most groups are stable at slightly larger than optimal sizes. Because it generally benefits an individual to join an optimally-sized group, despite slightly decreasing the advantage for all members, groups may continue to increase in size until it is more advantageous to remain alone than to join an overly full group.

Tinbergen's Four Questions for Ethologists

Niko Tinbergen argued that ethology always needed to include four kinds of explanation in any instance of behaviour:

- Function – How does the behaviour affect the animal's chances of survival and reproduction? Why does the animal respond that way instead of some other way?

- Causation – What are the stimuli that elicit the response, and how has it been modified by recent learning?

- Development – How does the behaviour change with age, and what early experiences are necessary for the animal to display the behaviour?

- Evolutionary history – How does the behaviour compare with similar behaviour in related species, and how might it have begun through the process of phylogeny?

These explanations are complementary rather than mutually exclusive—all instances of behaviour require an explanation at each of these four levels. For example, the function of eating is to acquire nutrients (which ultimately aids survival and reproduction), but the immediate cause of eating is hunger (causation). Hunger and eating are evolutionarily ancient and are found in many species (evolutionary history), and develop early within an organism's lifespan (development). It is easy to confuse such questions—for example, to argue that people eat because they're hungry and not to acquire nutrients—without realizing that the reason people experience hunger is because it causes them to acquire nutrients.

Arachnology

Arachnology is the scientific study of spiders and related animals such as scorpions, pseudoscorpions, and harvestmen, collectively called arachnids. Those who study spiders and other arachnids are arachnologists.

Nephila clavipes

Arachnology as a Science

Arachnologists are primarily responsible for classifying arachnids and studying aspects of their biology. In the popular imagination they are sometimes referred to as 'spider experts'. Disciplines within arachnology include naming species and determining their evolutionary relationships to one another (taxonomy and systematics), studying how they interact with other members of their species and/or their environment (behavioural ecology), or how they are distributed in different regions and habitats (faunistics). Other arachnologists perform research on the anatomy or physiology of arachnids, including the venom of spiders and scorpions. Others study the impact of spiders in agricultural ecosystems and whether they can be used as biological control agents.

Subdisciplines

Arachnology can be broken down into more specific specialties. These topics include:

- acarology - the study of ticks and mites.

- araneology - the study of spiders.

- scorpiology - the study of scorpions.

Arachnological Societies

Arachnologists are served by a number of scientific societies, both national and international in scope. Their main role is to encourage the exchange of ideas between researchers, to organise meetings and congresses, and in a number of cases to publish academic journals. Some are also involved in outreach programs, like the *European spider of the year*, which raise awareness of these animals among the general public.

- American Arachnological Society

- Iranian Arachnological Society

- British Arachnological Society

- International Society of Arachnology

Popular Arachnology

In the 1970s, arachnids - particularly tarantulas - started to become popular as exotic pets. Many tarantulas thus become more widely known by their common names such as the Mexican redknee tarantula (*Brachypelma smithi*).

Various societies now focus on the husbandry, care, study and captive breeding of tarantulas, and other arachnids. They also typically produce journals or newsletters with articles and advice on these subjects.

Entomology

Entomology is the scientific study of insects, a branch of zoology. In the past the term "insect" was more vague, and historically the definition of entomology included the study of terrestrial animals in other arthropod groups or other phyla, such as arachnids, myriapods, earthworms, land snails, and slugs. This wider meaning may still be encountered in informal use.

Like several of the other fields that are categorized within zoology, entomology is a taxon-based category; any form of scientific study in which there is a focus on insect-related inquiries is, by definition, entomology. Entomology therefore overlaps with a cross-section of topics as diverse as molecular genetics, behavior, biomechanics, biochemistry, systematics, physiology, developmental biology, ecology, morphology, paleontology, mathematics, anthropology, robotics, agriculture, nutrition, forensic science, and more.

A phasmid, mimicking a leaf

At some 1.3 million described species, insects account for more than two-thirds of all known organisms, date back some 400 million years, and have many kinds of interactions with humans and other forms of life on earth.

History

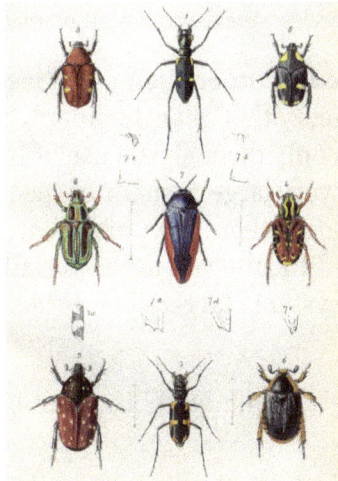

Plate from *Transactions of the Entomological Society*, 1848

Entomology is rooted in nearly all human cultures from prehistoric times, primarily in the context of agriculture (especially biological control and beekeeping), but scientific study began only as recently as the 16th century.

William Kirby is widely considered as the father of Entomology. In collaboration with William Spence, he published a definitive entomological encyclopedia, *Introduction to Entomology*, regarded as the subject's foundational text. He also helped to found the Royal Entomological Society in London in 1833, one of the earliest such societies in the world; earlier antecedents, such as the Aurelian society date back to the 1740s.

Entomology developed rapidly in the 19th and 20th centuries, and was studied by large numbers of people, including such notable figures as Charles Darwin, Jean-Henri Fabre, Vladimir Nabokov, Karl von Frisch (winner of the 1973 Nobel Prize in Physiology or Medicine), and two-time Pulitzer Prize winner E. O. Wilson.

Identification of Insects

These 100 *Trigonopterus* species were described simultaneously using DNA barcoding

Most insects can easily be recognized to order such as Hymenoptera (bees, wasps, and ants) or Coleoptera (beetles). However, insects other than Lepidoptera (butterflies and moths) are typically identifiable to genus or species only through the use of Identification keys and Monographs. Because the class Insecta contains a very large number of species (over 330,000 species of beetles alone) and the characteristics separating them are unfamiliar, and often subtle (or invisible without a microscope), this is often very difficult even for a specialist. This has led to the development of automated species identification systems targeted on insects, for example, Daisy, ABIS, SPIDA and Draw-wing

Insect identification is an increasingly common hobby, with butterflies and dragonflies being the most popular.

In Pest Control

In 1994 the Entomological Society of America launched a new professional certification program for the pest control industry called The Associate Certified Entomologist (ACE). To qualify as a "true entomologist" an individual would normally require an advanced degree, with most entomologists pursuing their PhD. While not true entomologists in the traditional sense, individuals who attain the ACE certification may be referred to as ACEs, Amateur entomologists, Associate entomologists or –more commonly– Associate-Certified Entomologists.

Taxonomic Specialization

Part of a large beetle collection

Many entomologists specialize in a single order or even a family of insects, and a number of these subspecialties are given their own informal names, typically (but not always) derived from the scientific name of the group:

- Coleopterology - beetles
- Dipterology - flies
- Hemipterology - true bugs
- Isopterology - termites
- Lepidopterology - moths and butterflies
- Melittology (or *Apiology*) - bees
- Myrmecology - ants
- Orthopterology - grasshoppers, crickets, etc.
- Trichopterology - caddis flies
- Vespology - Social wasps

Organizations

Like other scientific specialties, entomologists have a number of local, national, and international organizations. There are also many organizations specializing in specific subareas.

- Amateur Entomologists' Society
- Deutsches Entomologisches Institut
- Entomological Society of America
- Entomological Society of Canada

- Entomological Society of Japan
- International Union for the Study of Social Insects
- Netherlands Entomological Society
- Royal Belgian Entomological Society
- Royal Entomological Society of London
- Société entomologique de France

Museums

Here is a list of selected museums which contain very large insect collections.

Africa

- Natal Museum, Pietermaritzburg, South Africa

Europe

- Muséum national d'histoire naturelle, Paris, France
- Museum für Naturkunde, Berlin, Germany
- Natural History Museum, Budapest Hungarian Natural History Museum
- Natural History Museum, Geneva
- Natural History Museum, Leiden, the Netherlands
- Natural History Museum, London, United Kingdom
- Natural History Museum, Oslo Norway
- Natural History Museum, St. Petersburg Zoological Collection of the Russian Academy of Science
- Naturhistorisches Museum, Vienna, Austria
- Oxford University Museum of Natural History, Oxford
- Royal Museum for Central Africa, Brussels, Belgium
- Swedish Museum of Natural History, Stockholm, Sweden
- The Bavarian State Collection of Zoology Zoologische Staatssammlung München
- World Museum Liverpool, the *Bug House*

United States

- Academy of Natural Sciences of Philadelphia
- American Museum of Natural History, New York City

- Auburn University Museum of Natural History, Auburn, Alabama
- Audubon Insectarium, New Orleans
- Bohart Museum of Entomology, Davis, California
- California Academy of Sciences, San Francisco
- Carnegie Museum of Natural History, Pittsburgh
- Essig Museum, Berkeley, California
- Field Museum of Natural History, Chicago
- Florida Museum of Natural History, University of Florida, Gainesville, Florida
- Illinois Natural History Survey, Champaign, Illinois
- J. Gordon Edwards Museum, San Jose, California
- Museum of Comparative Zoology, Cambridge, Massachusetts
- Natural History Museum of Los Angeles County, Los Angeles
- National Museum of Natural History, Washington, D.C.
- New Mexico State University Arthropod Museum
- North Carolina State University Insect Museum, Raleigh, North Carolina
- Peabody Museum of Natural History, New Haven, Connecticut
- The National Museum of Play, Rochester, N.Y.
- Texas A&M University, College Station, Texas
- University of Minnesota, St. Paul campus (UMSP), Minnesota
- University of Kansas Natural History Museum, Lawrence, Kansas
- University of Nebraska State Museum, Lincoln, Nebraska
- University of Missouri Enns Entomology Museum, University of Missouri, Columbia, Missouri

Canada

- Canadian Museum of Nature, Ottawa
- Canadian National Collection of Insects, Arachnids and Nematodes, Ottawa, Ontario
- E.H. Strickland Entomological Museum, University of Alberta, Edmonton, Alberta
- Lyman Entomological Museum, Macdonald Campus of McGill University, Sainte-Anne-de-Bellevue, Quebec
- Montreal Insectarium, Montreal, Quebec

- Newfoundland Insectarium, Reidville, Newfoundland and Labrador

- Royal Alberta Museum, Edmonton, Alberta

- Royal Ontario Museum, Toronto

- University of Guelph Insect Collection, Guelph, Ontario

- Victoria Bug Zoo, Victoria, British Columbia

Anthropology

Anthropology is the study of various aspects of humans within past and present societies. Social anthropology and cultural anthropology study the norms and values of societies. Linguistic anthropology studies how language affects social life. Biological or physical anthropology studies the biological development of humans.

Archaeology, which studies past human cultures through investigation of physical evidence, is thought of as a branch of anthropology in the United States, while in Europe, it is viewed as a discipline in its own right, or grouped under other related disciplines such as history.

Origin and Development of the Term

The abstract noun *anthropology* is first attested in reference to history. Its present use first appeared in Renaissance Germany in the works of Magnus Hundt and Otto Casmann. (Its adjectival form appeared in the works of Aristotle.) It began to be used in English, possibly via French *anthropologie*, by the early 18th century.

Through The 19th Century

In 1647, the Bartholins, founders of the University of Copenhagen, defined *l'anthropologie* as follows:

Anthropology, that is to say the science that treats of man, is divided ordinarily and with reason into Anatomy, which considers the body and the parts, and Psychology, which speaks of the soul.

Sporadic use of the term for some of the subject matter occurred subsequently, such as the use by Étienne Serres in 1839 to describe the natural history, or paleontology, of man, based on comparative anatomy, and the creation of a chair in anthropology and ethnography in 1850 at the National Museum of Natural History (France) by Jean Louis Armand de Quatrefages de Bréau. Various short-lived organizations of anthropologists had already been formed. The Société Ethnologique de Paris, the first to use Ethnology, was formed in 1839. Its members were primarily anti-slavery activists. When slavery was abolished in France in 1848 the Société was abandoned.

Meanwhile, the Ethnological Society of New York, currently the American Ethnological Society, was founded on its model in 1842, as well as the Ethnological Society of London in 1843, a breakaway group of the Aborigines' Protection Society. These anthropologists of the times were liberal, anti-slavery, and pro-human-rights activists. They maintained international connections.

Anthropology and many other current fields are the intellectual results of the comparative methods developed in the earlier 19th century. Theorists in such diverse fields as anatomy, linguistics, and Ethnology, making feature-by-feature comparisons of their subject matters, were beginning to suspect that similarities between animals, languages, and folkways were the result of processes or laws unknown to them then. For them, the publication of Charles Darwin's *On the Origin of Species* was the epiphany of everything they had begun to suspect. Darwin himself arrived at his conclusions through comparison of species he had seen in agronomy and in the wild.

Darwin and Wallace unveiled evolution in the late 1850s. There was an immediate rush to bring it into the social sciences. Paul Broca in Paris was in the process of breaking away from the Société de biologie to form the first of the explicitly anthropological societies, the Société d'Anthropologie de Paris, meeting for the first time in Paris in 1859. When he read Darwin he became an immediate convert to *Transformisme*, as the French called evolutionism. His definition now became "the study of the human group, considered as a whole, in its details, and in relation to the rest of nature".

Broca, being what today would be called a neurosurgeon, had taken an interest in the pathology of speech. He wanted to localize the difference between man and the other animals, which appeared to reside in speech. He discovered the speech center of the human brain, today called Broca's area after him. His interest was mainly in Biological anthropology, but a German philosopher specializing in psychology, Theodor Waitz, took up the theme of general and social anthropology in his six-volume work, entitled *Die Anthropologie der Naturvölker*, 1859–1864. The title was soon translated as "The Anthropology of Primitive Peoples". The last two volumes were published posthumously.

Waitz defined anthropology as "the science of the nature of man". By nature he meant matter animated by "the Divine breath"; i.e., he was an animist. Following Broca's lead, Waitz points out that anthropology is a new field, which would gather material from other fields, but would differ from them in the use of comparative anatomy, physiology, and psychology to differentiate man from "the animals nearest to him". He stresses that the data of comparison must be empirical, gathered by experimentation. The history of civilization as well as ethnology are to be brought into the comparison. It is to be presumed fundamentally that the species, man, is a unity, and that "the same laws of thought are applicable to all men".

Waitz was influential among the British ethnologists. In 1863 the explorer Richard Francis Burton and the speech therapist James Hunt broke away from the Ethnological Society of London to form the Anthropological Society of London, which henceforward would follow the path of the new anthropology rather than just ethnology. It was the 2nd society dedicated to general anthropology in existence. Representatives from the French *Société* were present, though not Broca. In his keynote address, printed in the first volume of its new publication, *The Anthropological Review*, Hunt stressed the work of Waitz, adopting his definitions as a standard.[n 5] Among the first associates were the young Edward Burnett Tylor, inventor of cultural anthropology, and his brother Alfred Tylor, a geologist. Previously Edward had referred to himself as an ethnologist; subsequently, an anthropologist.

Similar organizations in other countries followed: The American Anthropological Association in 1902, the Anthropological Society of Madrid (1865), the Anthropological Society of Vienna (1870), the Italian Society of Anthropology and Ethnology (1871), and many others subsequently. The majority of these were evolutionist. One notable exception was the Berlin Society of Anthropology (1869) founded by Rudolph Virchow, known for his vituperative attacks on the evolutionists. Not religious himself, he insisted that Darwin's conclusions lacked empirical foundation.

During the last three decades of the 19th century a proliferation of anthropological societies and associations occurred, most independent, most publishing their own journals, and all international in membership and association. The major theorists belonged to these organizations. They supported the gradual osmosis of anthropology curricula into the major institutions of higher learning. By 1898 the American Association for the Advancement of Science was able to report that 48 educational institutions in 13 countries had some curriculum in anthropology. None of the 75 faculty members were under a department named anthropology.

20th and 21st Centuries

This meagre statistic expanded in the 20th century to comprise anthropology departments in the majority of the world's higher educational institutions, many thousands in number. Anthropology has diversified from a few major subdivisions to dozens more. Practical anthropology, the use of anthropological knowledge and technique to solve specific problems, has arrived; for example, the presence of buried victims might stimulate the use of a forensic archaeologist to recreate the final scene. Organization has reached global level. For example, the World Council of Anthropological Associations (WCAA), "a network of national, regional and international associations that aims to promote worldwide communication and cooperation in anthropology", currently contains members from about three dozen nations.

Since the work of Franz Boas and Bronisław Malinowski in the late 19th and early 20th centuries, *social* anthropology in Great Britain and *cultural* anthropology in the US have been distinguished from other social sciences by its emphasis on cross-cultural comparisons, long-term in-depth examination of context, and the importance it places on participant-observation or experiential immersion in the area of research. Cultural anthropology in particular has emphasized cultural relativism, holism, and the use of findings to frame cultural critiques. This has been particularly prominent in the United States, from Boas' arguments against 19th-century racial ideology, through Margaret Mead's advocacy for gender equality and sexual liberation, to current criticisms of post-colonial oppression and promotion of multiculturalism. Ethnography is one of its primary research designs as well as the text that is generated from anthropological fieldwork.

In Great Britain and the Commonwealth countries, the British tradition of social anthropology tends to dominate. In the United States, anthropology has traditionally been divided into the four field approach developed by Franz Boas in the early 20th century: *biological* or *physical* anthropology; *social*, *cultural*, or *sociocultural* anthropology; and archaeology; plus anthropological linguistics. These fields frequently overlap, but tend to use different methodologies and techniques.

European countries with overseas colonies tended to practice more ethnology (a term coined and defined by Adam F. Kollár in 1783). It is sometimes referred to as sociocultural anthropology in the parts of the world that were influenced by the European tradition.

Fields

Anthropology is a global discipline where humanities, social, and natural sciences are forced to confront one another. Anthropology builds upon knowledge from natural sciences, including the discoveries about the origin and evolution of *Homo sapiens*, human physical traits, human behavior, the variations among different groups of humans, how the evolutionary past of *Homo sapiens* has influenced its social organization and culture, and from social sciences, including the organization of human social and cultural relations, institutions, social conflicts, etc. Early anthropology originated in Classical Greece and Persia and studied and tried to understand observable cultural diversity. As such, anthropology has been central in the development of several new (late 20th century) interdisciplinary fields such as cognitive science, global studies, and various ethnic studies.

According to Clifford Geertz,

"anthropology is perhaps the last of the great nineteenth-century conglomerate disciplines still for the most part organizationally intact. Long after natural history, moral philosophy, philology, and political economy have dissolved into their specialized successors, it has remained a diffuse assemblage of ethnology, human biology, comparative linguistics, and prehistory, held together mainly by the vested interests, sunk costs, and administrative habits of academia, and by a romantic image of comprehensive scholarship."

Sociocultural anthropology has been heavily influenced by structuralist and postmodern theories, as well as a shift toward the analysis of modern societies. During the 1970s and 1990s, there was an epistemological shift away from the positivist traditions that had largely informed the discipline. During this shift, enduring questions about the nature and production of knowledge came to occupy a central place in cultural and social anthropology. In contrast, archaeology and biological anthropology remained largely positivist. Due to this difference in epistemology, the four sub-fields of anthropology have lacked cohesion over the last several decades.

Sociocultural

Sociocultural anthropology draws together the principle axes of cultural anthropology and social anthropology. Cultural anthropology is the comparative study of the manifold ways in which people *make sense* of the world around them, while social anthropology is the study of the *relationships* among persons and groups. Cultural anthropology is more related to philosophy, literature and the arts (how one's culture affects experience for self and group, contributing to more complete understanding of the people's knowledge, customs, and institutions), while social anthropology is more related to sociology and history. in that it helps develop understanding of social structures, typically of others and other populations (such as minorities, subgroups, dissidents, etc.). There is no hard-and-fast distinction between them, and these categories overlap to a considerable degree.

Inquiry in sociocultural anthropology is guided in part by cultural relativism, the attempt to understand other societies in terms of their own cultural symbols and values. Accepting other cultures in their own terms moderates reductionism in cross-cultural comparison. This project is often accommodated in the field of ethnography. Ethnography can refer to both a methodology and the product of ethnographic research, i.e. an ethnographic monograph. As methodology, ethnography is based upon long-term fieldwork within a community or other research site. Participant observation is one of the foundational methods of social and cultural anthropology. Ethnology involves the systematic comparison of different cultures. The process of participant-observation can be especially helpful to understanding a culture from an emic (conceptual, vs. etic, or technical) point of view.

The study of kinship and social organization is a central focus of sociocultural anthropology, as kinship is a human universal. Sociocultural anthropology also covers economic and political organization, law and conflict resolution, patterns of consumption and exchange, material culture, technology, infrastructure, gender relations, ethnicity, childrearing and socialization, religion, myth, symbols, values, etiquette, worldview, sports, music, nutrition, recreation, games, food, festivals, and language (which is also the object of study in linguistic anthropology).

Comparison across cultures is a key element of method in sociocultural anthropology, including the industrialized (and de-industrialized) West. Cultures in the Standard Cross-Cultural Sample (SCCS) of world societies are:

Biological

Forensic anthropologists can help identify skeletonized human remains, such as these found lying in scrub in Western Australia, c. 1900–1910.

Biological Anthropology and Physical Anthropology are synonymous terms to describe anthropological research focused on the study of humans and non-human primates in their biological, evolutionary, and demographic dimensions. It examines the biological and social factors that have affected the evolution of humans and other primates, and that generate, maintain or change contemporary genetic and physiological variation.

Archaeological

Archaeology is the study of the human past through its material remains. Artifacts, faunal remains, and human altered landscapes are evidence of the cultural and material lives of past societies. Archaeologists examine these material remains in order to deduce patterns of past human behavior and cultural practices. Ethnoarchaeology is a type of archaeology that studies the practices and material remains of living human groups in order to gain a better understanding of the evidence left behind by past human groups, who are presumed to have lived in similar ways.

Excavations at the 3800-year-old Edgewater Park Site, Iowa

Linguistic

Linguistic anthropology (also called anthropological linguistics) seeks to understand the processes of human communications, verbal and non-verbal, variation in language across time and space, the social uses of language, and the relationship between language and culture. It is the branch of anthropology that brings linguistic methods to bear on anthropological problems, linking the analysis of linguistic forms and processes to the interpretation of sociocultural processes. Linguistic anthropologists often draw on related fields including sociolinguistics, pragmatics, cognitive linguistics, semiotics, discourse analysis, and narrative analysis.

Art, Media, Music, Dance and Film

Art

One of the central problems in the anthropology of art concerns the universality of 'art' as a cultural phenomenon. Several anthropologists have noted that the Western categories of 'painting', 'sculpture', or 'literature', conceived as independent artistic activities, do not exist, or exist in a significantly different form, in most non-Western contexts. To surmount this difficulty, anthropologists of art have focused on formal features in objects which, without exclusively being 'artistic', have certain evident 'aesthetic' qualities. Boas' *Primitive Art*, Claude Lévi-Strauss' *The Way of the Masks* (1982) or Geertz's 'Art as Cultural System' (1983) are some examples in this trend to transform the anthropology of 'art' into an anthropology of culturally specific 'aesthetics'.

Media

A Punu tribe mask. Gabon Central Africa

Media anthropology (also known as anthropology of media or mass media) emphasizes ethnographic studies as a means of understanding producers, audiences, and other cultural and social aspects of mass media. The types of ethnographic contexts explored range from contexts of media production (e.g., ethnographies of newsrooms in newspapers, journalists in the field, film production) to contexts of media reception, following audiences in their everyday responses to media. Other types include cyber anthropology, a relatively new area of internet research, as well as ethnographies of other areas of research which happen to involve media, such as development work, social movements, or health education. This is in addition to many classic ethnographic contexts, where media such as radio, the press, new media and television have started to make their presences felt since the early 1990s.

Music

Ethnomusicology is an academic field encompassing various approaches to the study of music (broadly defined), that emphasize its cultural, social, material, cognitive, biological, and other dimensions or contexts instead of or in addition to its isolated sound component or any particular repertoire.

Visual

Visual anthropology is concerned, in part, with the study and production of ethnographic photography, film and, since the mid-1990s, new media. While the term is sometimes used interchangeably with ethnographic film, visual anthropology also encompasses the anthropological study of visual representation, including areas such as performance, museums, art, and the production and reception of mass media. Visual representations from all cultures, such as sandpaintings, tattoos, sculptures and reliefs, cave paintings, scrimshaw, jewelry, hieroglyphics, paintings and photographs are included in the focus of visual anthropology.

Economic, Political Economic, Applied and Development

Economic

Economic anthropology attempts to explain human economic behavior in its widest historic, geographic and cultural scope. It has a complex relationship with the discipline of economics, of which it is highly critical. Its origins as a sub-field of anthropology begin with the Polish-British founder of Anthropology, Bronislaw Malinowski, and his French compatriot, Marcel Mauss, on the nature of gift-giving exchange (or reciprocity) as an alternative to market exchange. Economic Anthropology remains, for the most part, focused upon exchange. The school of thought derived from Marx and known as Political Economy focuses on production, in contrast. Economic Anthropologists have abandoned the primitivist niche they were relegated to by economists, and have now turned to examine corporations, banks, and the global financial system from an anthropological perspective.

Political Economy

Political economy in anthropology is the application of the theories and methods of Historical Materialism to the traditional concerns of anthropology, including, but not limited to, non-capitalist societies. Political Economy introduced questions of history and colonialism to ahistorical anthropological theories of social structure and culture. Three main areas of interest rapidly developed. The first of these areas was concerned with the "pre-capitalist" societies that were subject to evolutionary "tribal" stereotypes. Sahlins work on Hunter-gatherers as the 'original affluent society' did much to dissipate that image. The second area was concerned with the vast majority of the world's population at the time, the peasantry, many of whom were involved in complex revolutionary wars such as in Vietnam. The third area was on colonialism, imperialism, and the creation of the capitalist world-system. More recently, these Political Economists have more directly addressed issues of industrial (and post-industrial) capitalism around the world.

Applied

Applied Anthropology refers to the application of the method and theory of anthropology to the analysis and solution of practical problems. It is a, "complex of related, research-based, instrumental methods which produce change or stability in specific cultural systems through the provision of data, initiation of direct action, and/or the formulation of policy". More simply, applied anthropology is the practical side of anthropological research; it includes researcher involvement and activism within the participating community. It is closely related to Development anthropology (distinct from the more critical Anthropology of development).

Development

Anthropology of development tends to view development from a *critical* perspective. The kind of issues addressed and implications for the approach simply involve pondering why, if a key development goal is to alleviate poverty, is poverty increasing? Why is there such a gap between plans and outcomes? Why are those working in development so willing to disregard history and the lessons it might offer? Why is development so externally driven rather than having an internal basis? In short why does so much planned development fail?

Kinship, Feminism, Gender and Sexuality

Kinship

Kinship can refer both to *the study of* the patterns of social relationships in one or more human cultures, or it can refer to *the patterns of social relationships* themselves. Over its history, anthropology has developed a number of related concepts and terms, such as "descent", "descent groups", "lineages", "affines", "cognates", and even "fictive kinship". Broadly, kinship patterns may be considered to include people related both by descent (one's social relations during development), and also relatives by marriage.

Feminist

Feminist anthropology is a four field approach to anthropology (archeological, biological, cultural, linguistic) that seeks to reduce male bias in research findings, anthropological hiring practices, and the scholarly production of knowledge. Anthropology engages often with feminists from non-Western traditions, whose perspectives and experiences can differ from those of white European and American feminists. Historically, such 'peripheral' perspectives have sometimes been marginalized and regarded as less valid or important than knowledge from the western world. Feminist anthropologists have claimed that their research helps to correct this systematic bias in mainstream feminist theory. Feminist anthropologists are centrally concerned with the construction of gender across societies. Feminist anthropology is inclusive of birth anthropology as a specialization.

Medical, Nutritional, Psychological, Cognitive and Transpersonal

Medical

Medical anthropology is an interdisciplinary field which studies "human health and disease, health care systems, and biocultural adaptation". Currently, research in medical anthropology is one of the main growth areas in the field of anthropology as a whole. It focuses on the following six basic fields:

- the development of systems of medical knowledge and medical care

- the patient-physician relationship

- the integration of alternative medical systems in culturally diverse environments

- the interaction of social, environmental and biological factors which influence health and illness both in the individual and the community as a whole

- the critical analysis of interaction between psychiatric services and migrant populations ("critical ethnopsychiatry": Beneduce 2004, 2007)

- the impact of biomedicine and biomedical technologies in non-Western settings

Other subjects that have become central to medical anthropology worldwide are violence and social suffering (Farmer, 1999, 2003; Beneduce, 2010) as well as other issues that involve physical and psychological harm and suffering that are not a result of illness. On the other hand, there are fields that intersect with medical anthropology in terms of research methodology and theoretical production, such as *cultural psychiatry* and *transcultural psychiatry* or *ethnopsychiatry*.

Nutritional

Nutritional anthropology is a synthetic concept that deals with the interplay between economic systems, nutritional status and food security, and how changes in the former affect the latter. If economic and environmental changes in a community affect access to food, food security, and dietary health, then this interplay between culture and biology is in turn connected to broader historical and economic trends associated with globalization. Nutritional status affects overall health status, work performance potential, and the overall potential for economic development (either in terms of human development or traditional western models) for any given group of people.

Psychological

Psychological anthropology is an interdisciplinary subfield of anthropology that studies the interaction of cultural and mental processes. This subfield tends to focus on ways in which humans' development and enculturation within a particular cultural group—with its own history, language, practices, and conceptual categories—shape processes of human cognition, emotion, perception, motivation, and mental health. It also examines how the understanding of cognition, emotion, motivation, and similar psychological processes inform or constrain our models of cultural and social processes.

Cognitive

Cognitive anthropology seeks to explain patterns of shared knowledge, cultural innovation, and transmission over time and space using the methods and theories of the cognitive sciences (especially experimental psychology and evolutionary biology) often through close collaboration with historians, ethnographers, archaeologists, linguists, musicologists and other specialists engaged in the description and interpretation of cultural forms. Cognitive anthropology is concerned with what people from different groups know and how that implicit knowledge changes the way people perceive and relate to the world around them.

Transpersonal

Transpersonal anthropology studies the relationship between altered states of consciousness and culture. As with transpersonal psychology, the field is much concerned with altered states of consciousness (ASC) and transpersonal experience. However, the field differs from mainstream transpersonal psychology in taking more cognizance of cross-cultural issues—for instance, the roles of myth, ritual, diet, and texts in evoking and interpreting extraordinary experiences.

Political and Legal

Political

Political anthropology concerns the structure of political systems, looked at from the basis of the structure of societies. Political anthropology developed as a discipline concerned primarily with politics in stateless societies, a new development started from the 1960s, and is still unfolding: anthropologists started increasingly to study more "complex" social settings in which the presence of states, bureaucracies and markets entered both ethnographic accounts and analysis of local phenomena. The turn towards complex societies meant that political themes were taken up at two main levels. First of all, anthropologists continued to study political organization and political phenomena that lay outside the state-regulated sphere (as in patron-client relations or tribal political organization). Second of all, anthropologists slowly started to develop a disciplinary concern with states and their institutions (and of course on the relationship between formal and informal political institutions). An anthropology of the state developed, and it is a most thriving field today. Geertz' comparative work on "Negara", the Balinese state is an early, famous example.

Legal

Legal anthropology or anthropology of law specializes in "the cross-cultural study of social ordering". Earlier legal anthropological research often focused more narrowly on conflict management, crime, sanctions, or formal regulation. More recent applications include issues such as human rights, legal pluralism, and political uprisings.

Public

Public Anthropology was created by Robert Borofsky, a professor at Hawaii Pacific University, to "demonstrate the ability of anthropology and anthropologists to effectively address problems beyond the discipline - illuminating larger social issues of our times as well as encouraging broad, public conversations about them with the explicit goal of fostering social change" (Borofsky 2004).

Nature, Science and Technology

Cyborg

Cyborg anthropology originated as a sub-focus group within the American Anthropological Association's annual meeting in 1993. The sub-group was very closely related to STS and the Society for the Social Studies of Science. Donna Haraway's 1985 *Cyborg Manifesto* could be considered the founding document of cyborg anthropology by first exploring the philosophical and sociological ramifications of the term. Cyborg anthropology studies humankind and its relations with the technological systems it has built, specifically modern technological systems that have reflexively shaped notions of what it means to be human beings.

Digital

Digital anthropology is the study of the relationship between humans and digital-era technology, and extends to various areas where anthropology and technology intersect. It is sometimes grouped with sociocultural anthropology, and sometimes considered part of material culture. The field is new, and thus has a variety of names with a variety of emphases. These include techno-anthropology, digital ethnography, cyberanthropology, and virtual anthropology.

Ecological

Ecological anthropology is defined as the "study of cultural adaptations to environments". The sub-field is also defined as, "the study of relationships between a population of humans and their biophysical environment". The focus of its research concerns "how cultural beliefs and practices helped human populations adapt to their environments, and how people used elements of their culture to maintain their ecosystems."

Environmental

Environmental anthropology is a sub-specialty within the field of anthropology that takes an active role in examining the relationships between humans and their environment across space and time. The contemporary perspective of environmental anthropology, and arguably at least the backdrop, if not the focus of most of the ethnographies and cultural fieldworks of today, is political ecology. Many characterize this new perspective as more informed with culture, politics and power, globalization, localized issues, and more. The focus and data interpretation is often used for arguments for/against or creation of policy, and to prevent corporate exploitation and damage of land. Often, the observer has become an active part of the struggle either directly (organizing, participation) or indirectly. Such is the case with environmental justice advocate Melissa Checker and her relationship with the people of Hyde Park.

Historical

Ethnohistory is the study of ethnographic cultures and indigenous customs by examining historical records. It is also the study of the history of various ethnic groups that may or may not exist today. Ethnohistory uses both historical and ethnographic data as its foundation. Its historical methods and materials go beyond the standard use of documents and manuscripts. Practitioners recognize the utility of such source material as maps, music, paintings, photography, folklore, oral tradition, site exploration, archaeological materials, museum collections, enduring customs, language, and place names.

Religion

The anthropology of religion involves the study of religious institutions in relation to other social institutions, and the comparison of religious beliefs and practices across cultures. Modern anthropology assumes that there is complete continuity between magical thinking and religion,[n 6] and that every religion is a cultural product, created by the human community that worships it.

Urban

Urban anthropology is concerned with issues of urbanization, poverty, and neoliberalism. Ulf Hannerz quotes a 1960s remark that traditional anthropologists were "a notoriously agoraphobic lot, anti-urban by definition". Various social processes in the Western World as well as in the "Third World" (the latter being the habitual focus of attention of anthropologists) brought the attention of "specialists in 'other cultures'" closer to their homes. There are two principle approaches in urban anthropology: by examining the types of cities or examining the social issues within the cities. These two methods are overlapping and dependent of each other. By defining different types of cities, one would use social factors as well as economic and political factors to categorize the cities. By directly looking at the different social issues, one would also be studying how they affect the dynamic of the city.

Key Topics by Field: Archaeological and Biological

Anthrozoology

Anthrozoology (also known as "human–animal studies") is the study of interaction between living things. It is a burgeoning interdisciplinary field that overlaps with a number of other disciplines, including anthropology, ethology, medicine, psychology, veterinary medicine and zoology. A major focus of anthrozoologic research is the quantifying of the positive effects of human-animal relationships on either party and the study of their interactions. It includes scholars from a diverse range of fields, including anthropology, sociology, biology, and philosophy.

Biocultural

Biocultural anthropology is the scientific exploration of the relationships between human biology and culture. Physical anthropologists throughout the first half of the 20th century viewed this relationship from a racial perspective; that is, from the assumption that typological human biological differences lead to cultural differences. After World War II the emphasis began to shift toward an effort to explore the role culture plays in shaping human biology.

Evolutionary

Evolutionary anthropology is the interdisciplinary study of the evolution of human physiology and human behaviour and the relation between hominins and non-hominin primates. Evolutionary anthropology is based in natural science and social science, combining the human development with socioeconomic factors. Evolutionary anthropology is concerned with both biological and cultural evolution of humans, past and present. It is based on a scientific approach, and brings together fields such as archaeology, behavioral ecology, psychology, primatology, and genetics. It is a dynamic and interdisciplinary field, drawing on many lines of evidence to understand the human experience, past and present.

Forensic

Forensic anthropology is the application of the science of physical anthropology and human osteology in a legal setting, most often in criminal cases where the victim's remains are in the advanced stages of decomposition. A forensic anthropologist can assist in the identification of deceased individuals whose remains are decomposed, burned, mutilated or otherwise unrecognizable. The adjective "forensic" refers to the application of this subfield of science to a court of law.

Palaeoanthropology

Paleoanthropology combines the disciplines of paleontology and physical anthropology. It is the study of ancient humans, as found in fossil hominid evidence such as petrifacted bones and footprints.

Organizations

Contemporary anthropology is an established science with academic departments at most universities and colleges. The single largest organization of Anthropologists is the American Anthropological Association (AAA), which was founded in 1903. Membership is made up of anthropologists from around the globe.

In 1989, a group of European and American scholars in the field of anthropology established the European Association of Social Anthropologists (EASA) which serves as a major professional organization for anthropologists working in Europe. The EASA seeks to advance the status of anthropology in Europe and to increase visibility of marginalized anthropological traditions and thereby contribute to the project of a global anthropology or world anthropology.

Hundreds of other organizations exist in the various sub-fields of anthropology, sometimes divided up by nation or region, and many anthropologists work with collaborators in other disciplines, such as geology, physics, zoology, paleontology, anatomy, music theory, art history, sociology and so on, belonging to professional societies in those disciplines as well.

List of Major Organizations

- American Anthropological Association
- American Ethnological Society
- Asociación de Antropólogos Iberoamericanos en Red, AIBR
- Moving Anthropology Student Network
- Anthropological Society of London
- Center for World Indigenous Studies
- Ethnological Society of London
- Institute of Anthropology and Ethnography
- Max Planck Institute for Evolutionary Anthropology
- Network of Concerned Anthropologists
- N. N. Miklukho-Maklai Institute of Ethnology and Anthropology
- Radical Anthropology Group
- Royal Anthropological Institute of Great Britain and Ireland
- Society for anthropological sciences
- Society for Applied Anthropology
- USC Center for Visual Anthropology

Controversial Ethical Stances

Anthropologists, like other researchers (especially historians and scientists engaged in field research), have over time assisted state policies and projects, especially colonialism.

Some commentators have contended:

- That the discipline grew out of colonialism, perhaps was in league with it, and derived some of its key notions from it, consciously or not.

- That ethnographic work was often ahistorical, writing about people as if they were "out of time" in an "ethnographic present" (Johannes Fabian, *Time and Its Other*).

Ethics of Cultural Relativism

As part of their quest for scientific objectivity, present-day anthropologists typically urge cultural relativism, which has an influence on all the sub-fields of anthropology. This is the notion that cultures should not be judged by another's values or viewpoints, but be examined dispassionately on their own terms. There should be no notions, in good anthropology, of one culture being better or worse than another culture.

Ethical commitments in anthropology include noticing and documenting genocide, infanticide, racism, mutilation (including circumcision and subincision), and torture. Topics like racism, slavery, and human sacrifice attract anthropological attention and theories ranging from nutritional deficiencies to genes to acculturation have been proposed, not to mention theories of colonialism and many others as root causes of Man's inhumanity to man. To illustrate the depth of an anthropological approach, one can take just one of these topics, such as "racism" and find thousands of anthropological references, stretching across all the major and minor sub-fields.

Ethical stance to Military Involvement

Anthropologists' involvement with the U.S. government, in particular, has caused bitter controversy within the discipline. Franz Boas publicly objected to US participation in World War I, and after the war he published a brief expose and condemnation of the participation of several American archaeologists in espionage in Mexico under their cover as scientists.

But by the 1940s, many of Boas' anthropologist contemporaries were active in the allied war effort against the "Axis" (Nazi Germany, Fascist Italy, and Imperial Japan). Many served in the armed forces, while others worked in intelligence (for example, Office of Strategic Services and the Office of War Information). At the same time, David H. Price's work on American anthropology during the Cold War provides detailed accounts of the pursuit and dismissal of several anthropologists from their jobs for communist sympathies.

Attempts to accuse anthropologists of complicity with the CIA and government intelligence activities during the Vietnam War years have turned up surprisingly little (although anthropologist Hugo Nutini was active in the stillborn Project Camelot). Many anthropologists (students and teachers) were active in the antiwar movement. Numerous resolutions condemning the war in all its aspects were passed overwhelmingly at the annual meetings of the American Anthropological Association (AAA).

Professional anthropological bodies often object to the use of anthropology for the benefit of the state. Their codes of ethics or statements may proscribe anthropologists from giving secret briefings. The Association of Social Anthropologists of the UK and Commonwealth (ASA) has called certain scholarship ethically dangerous. The AAA's current 'Statement of Professional Responsibility' clearly states that "in relation with their own government and with host governments ... no secret research, no secret reports or debriefings of any kind should be agreed to or given."

Anthropologists, along with other social scientists, are working with the US military as part of the US Army's strategy in Afghanistan. The *Christian Science Monitor* reports that "Counterinsurgency efforts focus on better grasping and meeting local needs" in Afghanistan, under the *Human Terrain System* (HTS) program; in addition, HTS teams are working with the US military in Iraq. In 2009, the American Anthropological Association's Commission on the Engagement of Anthropology with the US Security and Intelligence Communities released its final report concluding, in part, that, "When ethnographic investigation is determined by military missions, not subject to external review, where data collection occurs in the context of war, integrated into the goals of counterinsurgency, and in a potentially coercive environment – all characteristic factors of the HTS concept and its application – it can no longer be considered a legitimate professional exercise of anthropology. In summary, while we stress that constructive engagement between anthropology and the military is possible, CEAUSSIC suggests that the AAA emphasize the incompatibility of HTS with disciplinary ethics and practice for job seekers and that it further recognize the problem of allowing HTS to define the meaning of "anthropology" within DoD."

Post–World War II Developments

Before WWII British 'social anthropology' and American 'cultural anthropology' were still distinct traditions. After the war, enough British and American anthropologists borrowed ideas and methodological approaches from one another that some began to speak of them collectively as 'sociocultural' anthropology.

Basic Trends

There are several characteristics that tend to unite anthropological work. One of the central characteristics is that anthropology tends to provide a comparatively more holistic account of phenomena and tends to be highly empirical. The quest for holism leads most anthropologists to study a particular place, problem or phenomenon in detail, using a variety of methods, over a more extensive period than normal in many parts of academia.

In the 1990s and 2000s (decade), calls for clarification of what constitutes a culture, of how an observer knows where his or her own culture ends and another begins, and other crucial topics in writing anthropology were heard. These dynamic relationships, between what can be observed on the ground, as opposed to what can be observed by compiling many local observations remain fundamental in any kind of anthropology, whether cultural, biological, linguistic or archaeological.

Biological anthropologists are interested in both human variation and in the possibility of human universals (behaviors, ideas or concepts shared by virtually all human cultures). They use many different methods of study, but modern population genetics, participant observation and other techniques often take anthropologists "into the field," which means traveling to a community in its own setting, to do something called "fieldwork." On the biological or physical side, human measurements, genetic samples, nutritional data may be gathered and published as articles or monographs.

Along with dividing up their project by theoretical emphasis, anthropologists typically divide the world up into relevant time periods and geographic regions. Human time on Earth is divided up into relevant cultural traditions based on material, such as the Paleolithic and the Neolithic, of particular use in archaeology. Further cultural subdivisions according to tool types, such as Olduwan or Mousterian or Levalloisian help archaeologists and other anthropologists in understanding major trends in the human past. Anthropologists and geographers share approaches to Culture regions as well, since mapping cultures is central to both sciences. By making comparisons across cultural traditions (time-based) and cultural regions (space-based), anthropologists have developed various kinds of comparative method, a central part of their science.

Commonalities Between Fields

Because anthropology developed from so many different enterprises, including but not limited to fossil-hunting, exploring, documentary film-making, paleontology, primatology, antiquity dealings and curatorship, philology, etymology, genetics, regional analysis, ethnology, history, philosophy, and religious studies, it is difficult to characterize the entire field in a brief article, although attempts to write histories of the entire field have been made.

Some authors argue that anthropology originated and developed as the study of "other cultures", both in terms of time (past societies) and space (non-European/non-Western societies). For example, the classic of urban anthropology, Ulf Hannerz in the introduction to his seminal *Exploring the City: Inquiries Toward an Urban Anthropology* mentions that the "Third World" had habitually received most of attention; anthropologists who traditionally specialized in "other cultures" looked for them far away and started to look "across the tracks" only in late 1960s.

Now there exist many works focusing on peoples and topics very close to the author's "home". It is also argued that other fields of study, like History and Sociology, on the contrary focus disproportionately on the West.

In France, the study of Western societies has been traditionally left to sociologists, but this is increasingly changing, starting in the 1970s from scholars like Isac Chiva and journals like *Terrain* ("fieldwork"), and developing with the center founded by Marc Augé (*Le Centre d'anthropologie des mondes contemporains*, the Anthropological Research Center of Contemporary Societies).

Since the 1980s it has become common for social and cultural anthropologists to set ethnographic research in the North Atlantic region, frequently examining the connections between locations rather than limiting research to a single locale. There has also been a related shift toward broadening the focus beyond the daily life of ordinary people; increasingly, research is set in settings such as scientific laboratories, social movements, governmental and nongovernmental organizations and businesses.

Anthrozoology

Anthrozoology (also known as human–non-human-animal studies, or HAS) is the subset of ethno-biology that deals with interactions between humans and other animals. It is an interdisciplinary field that overlaps with other disciplines including anthropology, ethnology, medicine, psychology, veterinary medicine and zoology. A major focus of anthrozoologic research is the quantifying of the positive effects of human-animal relationships on either party and the study of their interactions. It includes scholars from fields such as anthropology, sociology, biology, history and philosophy.

Man's best friend: dogsled racing in Alaska

Anthrozoology scholars, such as Pauleen Bennett recognize the lack of scholarly attention given to non-human animals in the past, and to the relationships between human and non-human animals, especially in the light of the magnitude of animal representations, symbols, stories and their actual physical presence in human societies. Rather than a unified approach, the field currently consists of several methods adapted from the several participating disciplines to encompass human-non-human animal relationships and occasional efforts to develop *sui generis* methods.

Areas of Study

- The interaction and enhancement within captive animal interactions.
- Affective (emotional) or relational bonds between humans and animals
- Human perceptions and beliefs in respect of other animals
- How some animals fit into human societies
- How these vary between cultures, and change over times
- The study of animal domestication: how and why domestic animals evolved from wild species (paleoanthrozoology)

- Captive zoo animal bonds with keepers

- The social construction of animals and what it means to be animal

- The zoological gaze

- The human-animal bond

- Parallels between human-animal interactions and human-technology interactions

- The symbolism of animals in literature and art

- The history of animal domestication

- The intersections of speciesism, racism, and sexism

- The place of animals in human-occupied spaces

- The religious significance of animals throughout human history

- Exploring the cross-cultural ethical treatment of animals

- The critical evaluation of animal abuse and exploitation

- Mind, self, and personhood in nonhuman animals

Growth of The Field

There are currently 23 college programs in HAS or a related field in the United States, Canada, Great Britain, Germany, Israel and the Netherlands, as well as an additional eight veterinary school programs in North America, and over thirty HAS organizations in the US, Canada, Great Britain, Australia, France, Germany, New Zealand, Israel, Sweden, and Switzerland.

In the UK, the University of Exeter runs an MA in Anthrozoology which explores human-animal interactions from anthropological (cross-cultural) perspectives. Human animal interactions (HAI) involving companion animals are also studied by the Waltham Centre for Pet Nutrition, which partners with the US National Institutes of Health to research HAI in relation to child development and aging.

There are now three primary lists for HAS scholars and students—H-Animal, the Human-Animal Studies listserv, and NILAS, as well as the Critical Animal Studies list.

There are now over a dozen journals covering HAS issues, many of them founded in the last decade, and hundreds of HAS books, most of them published in the last decade. Brill, Berg, Johns Hopkins, Purdue, Columbia, Reaktion, Palgrave-McMillan, University of Minnesota, University of Illinois, and Oxford all offer either a HAS series or a large number of HAS books.

In addition, in 2006, Animals and Society Institute (ASI) began hosting the Human-Animal Studies Fellowship, a six-week program in which pre- and post-doctoral scholars work on a HAS research project at a university under the guidance of host scholars and distance peer scholars. Beginning in 2011, ASI has partnered with Wesleyan Animal Studies, who will be hosting the fellowship in conjunction with ASI. There are also a handful of HAS conferences per year, including those organized by ISAZ and NILAS, and the Minding Animals conference, held in 2009 in Australia. Finally, there are more HAS courses being taught now than ever before. The ASI website lists over 300 courses (primarily in North America, but also including Great Britain, New Zealand, Australia, Germany, and Poland) in twenty-nine disciplines at over 200 colleges and universities, not including over 100 law school courses.

Primatology

Olive baboon

Primatology is the scientific study of primates. It is a diverse discipline and researchers can be found in academic departments of anatomy, anthropology, biology, medicine, psychology, veterinary sciences and zoology, as well as in animal sanctuaries, biomedical research facilities, museums and zoos. Primatologists study both living and extinct primates in their natural habitats and in laboratories by conducting field studies and experiments in order to understand aspects of their evolution and behaviour. Primatologists often divide primates into three groups for study: dominant females, females and young, and peripheral males.

Sub-disciplines

As a science, primatology has many different sub-disciplines which vary in terms of theoretical and methodological approaches to the subject used in researching extant primates and their extinct ancestors.

There are two main centers of primatology, Western primatology and Japanese primatology. These two divergent disciplines stem from their unique cultural backgrounds and philosophies that went into their founding. Although, fundamentally, both Western and Japanese primatology share many of the same principles, the areas of their focus in primate research and their methods of obtaining data differ widely.

Western Primatology

Origins

Western primatology stems primarily from research by North American and European scientists. Early primate study focused primarily in medical research, but some scientists also conducted "civilizing" experiments on chimpanzees in order to gauge both primate intelligence and the limits of their brainpower...

Theory

The study of primatology looks at the biological and psychological aspects of non-human primates. The focus is on studying the common links between humans and primates. It is believed that by understanding our closest animal relatives, we might better understand the nature shared with our ancestors.

Methods

Primatology is a science. The general belief is that the scientific observation of nature must be either extremely limited, or completely controlled. Either way, the observers must be neutral to their subjects. This allows for data to be unbiased and for the subjects to be uninfluenced by human interference.

There are three methodological approaches in primatology: field study, the more realistic approach, laboratory study, the more controlled approach, and semi-free ranging, where primate habitat and wild social structure is replicated in a captive setting.

Field is done in natural environments, in which scientific observers watch primates in their natural habitat.

Laboratory study is done in controlled lab settings. In lab settings, scientists are able to perform controlled experimentation on the learning capabilities and behavioral patterns of the animals.

In semi-free ranging studies, scientists are able to watch how primates might act in the wild but have easier access to them, and the ability to control their environments. Such facilities include the Living Links Center at the Yerkes National Primate Research Center in Georgia and the Elgin Center at Lion Country Safari in Florida.

All types of primate study in the Western methodology are meant to be neutral. Although there are certain Western primatologists who do more subjective research, the emphasis in this discipline is on the objective.

Early field primatology tended to focus on individual researchers. Researchers such as Dian Fossey and Jane Goodall are examples of this. Long-term sites of research tend to be best associated with their founders, and this led to some tension between younger primatologists and the veterans in the field.

Notable Western Primatologists

- Nina Amstrup

- Jeanne Altmann
- Jean Baulu
- Irwin Bernstein
- Michelle Bezanson
- Sarah Blaffer Hrdy
- Christophe Boesch
- Sue Boinski
- Geoffrey Bourne
- Josep Call
- C. R. Carpenter
- Colin Chapman
- Dorothy Cheney
- Marina Cords
- Thomas Defler
- Frans de Waal
- Alan F. Dixson
- Joseph Erwin
- Alejandro Estrada
- Linda Fedigan
- Dian Fossey
- Dorothy Fragaszy
- Agustin Fuentes
- Birutė Galdikas
- Paul Garber
- Thomas R. Gillespie
- Jane Goodall
- Colin Groves
- Alexander Harcourt
- Harry Harlow
- Philip Hershkovitz

- William C. Osman Hill
- Gottfried Hohmann
- Lynne Isbell
- Alison Jolly
- Hans Kummer
- Boris Lapin
- Nadezhda Ladygina-Kohts
- Louis Leakey
- Donald Lindburg
- Robert D. Martin
- William Mason
- Bill McGrew
- Emil Wolfgang Menzel, Jr.
- Gary D. Mitchell
- John Mitani
- Russell Mittermeier
- John R. Napier
- Prudence H. Napier
- Nicholas Newton-Fisher
- William C. Osman Hill
- Matthew Richardson
- Shawn Ridgeway
- Anne E. Russon
- Anthony Rylands
- Jordi Sabater Pi
- Gene P. Sackett
- Robert Sapolsky
- Carel van Schaik
- Adolph H. Schultz
- Robert M. Seyfarth

- Joan Silk
- Meredith Small
- Barbara Smuts
- Charles Southwick
- Craig Stanford
- Karen Strier
- Tom Struhsaker
- Robert W. Sussman
- Michael Tomasello
- Omar Wasow
- Sherwood Washburn
- David P. Watts
- Barbara J. Welker
- Richard Wrangham
- Gabriel Zunino
- Susan Emily Perry
- Carlos A. Peres

Japanese Primatology

Origins

The discipline of Japanese primatology was developed out of animal ecology. It is mainly credited to Kinji Imanishi and Junichiro Itani. Imanishi was an animal ecologist who began studying wild horses before focusing more on primate ecology. He helped found the Primate Research Group in 1950. Junichiro was a renowned anthropologist and a professor at Kyoto University. He is a co-founder of the Primate Research Institute and the Centre for African Area Studies.

Theory

The Japanese discipline of primatology tends to be more interested in the social aspects of primates. Social evolution and anthropology are of primary interest to them. The Japanese theory believes that studying primates will give us insight into the duality of human nature: individual self vs. social self.

The traditional and cultural aspects of Japanese science lend themselves to an "older sibling" mentality. It is believed that animals should be treated with respect, but also a firm authority. This is not to say that the Japanese study of primatology is cruel – far from it – just that it does not feel that their subjects should be given reverential treatment.

One particular Japanese primatologist, Kawai Masao, introduced the concept of *kyokan*. This was the theory that the only way to attain reliable scientific knowledge was to attain a mutual relation, personal attachment and shared life with the animal subjects. Though Kawai is the only Japanese primatologist associated with the use of this term, the underlying principle is part of the foundation of Japanese primate research.

Methods

Japanese primatology is a carefully disciplined subjective science. It is believed that the best data comes through identification with your subject. Neutrality is eschewed in favour of a more casual atmosphere, where researcher and subject can mingle more freely. Domestication of nature is not only desirable, but necessary for study.

Japanese primatologists are renowned for their ability to recognise animals by sight, and indeed most primates in a research group are usually named and numbered. Comprehensive data on every single subject in a group is uniquely Japanese trait of primate research. Each member of the primate community has a part to play, and the Japanese researchers are interested in this complex interaction.

For Japanese researchers in primatology, the findings of the team are emphasised over the individual. The study of primates is a group effort, and the group will get the credit for it. A team of researchers may observe a group of primates for several years in order to gather very detailed demographic and social histories.

Notable Japanese Primatologists

- Imanishi, Kinji
- Junichiro, Itani
- Masao, Kawai
- Tetsuro Matsuzawa
- Toshisada Nishida
- Hiraiwa-Hasegawa, Mariko
- Jensen, Erik
- Satsue Mito

Primatology in Sociobiology

Where sociobiology attempts to understand the actions of all animal species within the context of advantageous and disadvantageous behaviors, primatology takes an exclusive look at the order Primates, which includes *Homo sapiens*. The interface between primatology and sociobiology examines in detail the evolution of primate behavioral processes, and what studying our closest living primate relatives can tell about our own minds. As the American anthropologist Earnest Albert Hooton used to say: "Primas sum: primatum nil a me alienum puto" ("I am a primate; nothing about primates is outside of my bailiwick"). The meeting point of these two disciplines has become a nexus of discussion on key issues concerning the evolution of sociality, the development and purpose of language and deceit, and the development and propagation of culture.

Additionally, this interface is of particular interest to the science watchers in science and technology studies, who examine the social conditions which incite, mould, and eventually react to scientific discoveries and knowledge. The STS approach to primatology and sociobiology stretches beyond studying the apes, into the realm of observing the people studying the apes.

Taxonomic Basis

Before Darwin, and before molecular biology, the father of modern taxonomy, Carl Linnaeus, organized natural objects into kinds, that we now know reflect their evolutionary relatedness. He sorted these kinds by morphology, the shape of the object. Animals such as monkeys, chimpanzees and orangutans resemble humans closely, so Linnaeus placed *Homo sapiens* together with other similar-looking organisms into the taxonomic order *Primates*. Modern molecular biology reinforced humanity's place within the Primate order. Humans and simians share the vast majority of their DNA, with chimpanzees sharing between 97-99% genetic identity with humans.

From Grooming to Speaking

Although social grooming is observed in many animal species, the grooming activities undertaken by primates are not strictly for the elimination of parasites. In primates, grooming is a social activity that strengthens relationships. The amount of grooming taking place between members of a troop is a potent indicator of alliance formation or troop solidarity. Robin Dunbar suggests a link between primate grooming and the development of human language. The size of the neocortex in a primate's brain correlates directly to the number of individuals it can keep track of socially, be it a troop of chimps or a tribe of humans.

This number is referred to as the monkeysphere. If a population exceeds the size outlined by its cognitive limitations, the group undergoes a schism. Set into an evolutionary context, the Dunbar number shows a drive for the development of a method of bonding that is less labor-intensive than grooming: language. As the monkeysphere grows, the amount of time that would need to be spent grooming troopmates soon becomes unmanageable. Furthermore, it is only possible to bond with one troopmate at a time while grooming. The evolution of vocal communication solves both the time constraint and the one-on-one problem, but at a price.

Language allows for bonding with multiple people at the same time at a distance, but the bonding produced by language is less intense. This view of language evolution covers the general biological trends needed for language development, but it takes another hypothesis to uncover the evolution of the cognitive processes necessary for language.

Modularity of the Primate Mind

Noam Chomsky's concept of innate language addresses the existence of universal grammar, which suggests a special kind of "device" all humans are born with whose sole purpose is language. Fodor's modular mind hypothesis expands on this concept, suggesting the existence of preprogrammed modules for dealing with many, or all aspects of cognition. Although these modules do not need to be physically distinct, they must be functionally distinct. Orangutans are currently being taught language at the Smithsonian National Zoo using a computer system developed by primatologist Dr. Francine Neago in conjunction with IBM.

The massive modularity theory thesis posits that there is a huge number of tremendously interlinked but specialized modules running programs called Darwinian algorithms, or DA. DA can be selected for just as a gene can, eventually improving cognition. The contrary theory, of generalist mind, suggests that the brain is just a big computer that runs one program, the mind. If the mind is a general computer, for instance, the ability to use reasoning should be identical regardless of the context. This is not what is observed. When faced with abstract numbers and letters with no "real world" significance, respondents of the Wason card test generally do very poorly. However, when exposed to a test with an identical rule set but socially relevant content, respondents score markedly higher. The difference is especially pronounced when the content is about reward and payment. This test strongly suggests that human logic is based on a module originally developed in a social environment to root out cheaters, and that either the module is at a huge disadvantage where abstract thinking is involved, or that other less effective modules are used when faced with abstract logic.

Further evidence supporting the modular mind has steadily emerged with some startling revelations concerning primates. A very recent study indicated that human babies and grown monkeys approach and process numbers in a similar fashion, suggesting an evolved set of DA for mathematics (Jordan). The conceptualization of both human infants and primate adults is cross-sensory, meaning that they can add 15 red dots to 20 beeps and approximate the answer to be 35 grey squares. As more evidence of basic cognitive modules are uncovered, they will undoubtedly form a more solid foundation upon which the more complex behaviors can be understood.

In contradiction to this, neuroscientist Jaak Panksepp has argued that the mind is not a computer nor is it massively modular. He states that no evidence of massive modularity or the brain as a digital computer has been gained through actual neuroscience, as opposed to psychological studies. He criticises psychologists who use the massive modularity thesis for not integrating neuroscience into their understanding.

The Primate Theory of Mind

Primate behavior, like human behavior, is highly social and ripe with the intrigue of kingmaking, powerplays, deception, cuckoldry, and apology. In order to understand the staggeringly complex nature of primate interactions, we look to theory of mind. Theory of mind asks whether or not an individual recognizes and can keep track of information asymmetry amongst individuals in the group, and whether or not they can attribute folk psychological states to their peers. If some primates can tell what others know and want and act accordingly, they can gain advantage and status.

Recently, chimpanzee theory of mind has been advanced by Felix Warneken of the Max Planck Institute. His studies have shown that chimpanzees can recognize whether a researcher desires a dropped object, and act accordingly by picking it up. Even more compelling is the observation that chimps will only act if the object is dropped in an accidental-looking manner: if the researcher drops the object in a way that appears intentional, the chimp will ignore the object.

In a related experiment, groups of chimps were given rope-pulling problems they could not solve individually. Warneken's subjects rapidly figured out which individual in the group was the best rope puller and assigned it the bulk of the task. This research is highly indicative of the ability of chimps to detect the folk psychological state of "desire", as well as the ability to recognize that other individuals are better at certain tasks than they are.

However primates do not always fare so well in situations requiring theory of mind. In one experiment pairs of chimpanzees who had been close grooming partners were offered two levers. Pressing one lever would bring them food and another would bring their grooming partner food. Pressing the lever to clearly give their grooming partner much-wanted food would not take away from how much food they themselves got. For some reason, the chimps were unwilling to depress the lever that would give their long-time chums food. It is plausible but unlikely that the chimps figured there was finite food and it would eventually decrease their own food reward. The experiments are open to such interpretations making it hard to establish anything for certain.

One phenomenon which would indicate a possible fragility of theory of mind in primates occurs when a baboon gets lost. Under such circumstances, the lost baboon generally makes "call barks" to announce that it is lost. Previous to the 1990s it was thought that these call barks would then be returned by the other baboons, similar to the case is in vervet monkeys. However, when researchers studied this formally in the past few years they found something surprising: Only the baboons who were lost would ever give call barks. Even if an infant was wailing in agony just a few hundred meters away, its mother who would clearly recognise its voice and would be frantic about his safety (or alternatively run towards her infant depending on her own perceived safety), would often simply stare in his direction visibly agitated. If the anguishing baboon mother made any type of call at all, the infant would instantly recognise her and run to her position. This type of logic appears to be lost on the baboon, suggesting a serious gap in theory of mind of this otherwise seemingly very intelligent primate species. However, it is also possible that baboons do not return call barks for ecological reasons, for example because returning the call bark might call attention to the lost baboon, putting it at greater risk from predators.

Criticisms

Scientific studies concerning primate and human behavior have been subject to the same set of political and social complications, or biases, as every other scientific discipline. The borderline and multidisciplinary nature of primatology and sociobiology make them ripe fields of study because they are amalgams of objective and subjective sciences. Current scientific practice, especially in the hard sciences, requires a total dissociation of personal experience from the finished scientific product (Bauchspies 8). This is a strategy that is incompatible with observational field studies, and weakens them in the eyes of hard science. As mentioned above, the Western school of primatology tries to minimize subjectivity, while the Japanese school of primatology tends to embrace the closeness inherent in studying nature.

Social critics of science, some operating from within the field, are critical of primatology and sociobiology. Claims are made that researchers bring pre-existing opinions on issues concerning human sociality to their studies, and then seek evidence that agrees with their worldview or otherwise furthers a sociopolitical agenda. In particular, the use of primatological studies to assert gender roles, and to both promote and subvert feminism has been a point of contention.

An example of this is Zuckerman's 1932 study of captive hamadryas baboons, as critiqued in Sturm and Fedigan's Changing Views on Primate Societies. Zuckerman observed male baboons kill each other in a captive environment. Whether intended or not, the study served to reinforce images of the male as the sole competitor in an often violent race to secure dominance and access to a harem of females. Despite unrealistic overcrowding and unnatural male-to-female ratios, Zuckerman's paper was viewed as good science at the time. These ideas were used to justify the status quo of human male dominance, and these interpretations were widely accepted and assumed to be the basis of a primate-wide template for behavior, including that of humans. Incidentally, the hamadryas baboon females are among the most submissive and most gender-unequal of all primates, although primates and humans share a tremendous variation in troop structure (Hrdy 101, Stone).

Several research papers on primate cognition were retracted in 2010. Their lead author, primatologist Marc Hauser, was dismissed from Harvard University after an internal investigation found evidence of scientific misconduct in his laboratory. Data supporting the authors' conclusion that cottontop tamarin monkeys displayed pattern-learning behavior similar to human infants reportedly could not be located after a three-year investigation.

Women in Primatology

Women receive the majority of Ph.D's in primatology. Londa Schiebinger, writing in 2001, estimated that women made up 80 percent of graduate students pursuing Ph.D's in primatology, up from 50 percent in the 1970s. Because of the high number of women, Schiebinger has even asserted that "Primatology is widely celebrated as a feminist science".

Changing Stereotypes

With attention to Darwin's perception about sexual selection, it was perceived that sexual selection acted differently on females and males. Early research emphasized male-male competition for females. It is widely believed that males tend to woo females, and that females were passive. For years this was the dominant interpretation, emphasizing competition among dominant males who controlled territorial boundaries and maintained order among lesser males. Females on the other hand were described as "dedicated mothers to small infants and sexually available to males in order of the males' dominance rank". Female-female competition was ignored. Schiebinger proposed that the failure to acknowledge female-female competitions could "skew notions of sexual selection" to "ignore interactions between males and females that go beyond the strict interpretation of sex as for reproduction only." In the 1960s primatologists started looking at what females did, slowly changing the stereotype of the passive female. We now know that females are active participants, and even leaders, within their groups. For instance, Rowell found that female baboons determine the route for daily foraging (primary ref needed). Similarly, Shirley Strum found that male investment in special relationships with females had greater productive payoff in comparison to a male's rank in a dominance hierarchy (primary ref needed). This emerging "female point of view" resulted in a reanalysis of how aggression, reproductive access, and dominance affect primate societies.

Schiebinger has also accused sociobiologists of producing the "corporate primate", described as "female baboons with briefcases, strategically competitive and aggressive." This contrasts with the notion that only men are competitive and aggressive. Observations have repeatedly demonstrated that female apes and monkeys also form stable dominance hierarchies and alliances with their male counterparts. Females display aggression, exercise sexual choice, and compete for resources, mates and territory, like their male counterparts.

Six different features of feminist science that characterize contemporary primatology (Fedigan).

1. Reflexivity: sensitivity to context and cultural bias in scientific work.

2. "The female point of view"

3. Respect for nature and an ethic cooperation with nature

4. Move away from reductionism

5. Promote humanitarian values rather than national interests

6. Diverse community, accessible and egalitarian

Schiebinger suggests that only two out of the six features characteristic of feminism. One of them is the discussion of the politics of participation and the attention placed on females as subjects of research.

The Evolution of Primatology

In 1970 Jeanne Altmann drew attention to representative sampling methods in which all individuals, not just the dominant and the powerful, were observed for equal periods of time. Prior to 1970, primatologists used "opportunistic sampling," which only recorded what caught their attention. Sarah Hrdy, a self-identified feminist, was among the first to apply what became known as sociobiological theory to primates. In her studies, she focuses on the need for females to win from males parental care for their offspring. Linda Fedigan views herself as a reporter or translator, working at the intersection between gender studies of science and the mainstream study of primatology. While some influential women challenged fundamental paradigms, Schiebinger suggests that science is constituted by numerous factors varying from gender roles and domestic issues that surround race and class to economic relations between researchers from Developed World countries and the Developing World countries in which most nonhuman primates reside.

Mammalogy

Siberian tiger

In zoology, mammalogy is the study of mammals – a class of vertebrates with characteristics such as homeothermic metabolism, fur, four-chambered hearts, and complex nervous systems. Mammalogy has also been known as "mastology," "theriology," and "therology." There are about 4,200 different species of animals which are considered mammals. The major branches to study in the mammalogy career include natural history, taxonomy and systematics, anatomy and physiology, ethology, ecology, and management and control. The approximate salary of a mammalogist varies from $20,000 to $60,000 a year, depending on their experience. Mammalogists are typically involved in activities such as conducting research, managing personnel, and writing proposals.

Mammalogy branches off into other taxonomically-oriented disciplines such as primatology (study of primates), and cetology (study of cetaceans).

A Career in Mammalogy

Students seeking a career in mammalogy may be disappointed to know that there are more mammalogists seeking employment than there are positions and this doesn't look like it will change in the foreseeable future. The competition surrounding open positions will be intense which behooves mammalogists to obtain the best training possible. Summer employment or volunteer work can be extremely valuable in this type of situation. Most positions in mammalogy will require a broad undergraduate background with the coursework depending on the specific field of mammalogy.

Ornithology

Ornithology is a branch of zoology that concerns the study of birds. Several aspects of ornithology differ from related disciplines, due partly to the high visibility and the aesthetic appeal of birds. Most marked among these is the extent of studies undertaken by amateurs working within the parameters of strict scientific methodology.

House sparrow (*Passer domesticus*)

The science of ornithology has a long history and studies on birds have helped develop several key concepts in evolution, behaviour and ecology such as the definition of species, the process of speciation, instinct, learning, ecological niches, guilds, island biogeography, phylogeography and conservation. While early ornithology was principally concerned with descriptions and distributions of species, ornithologists today seek answers to very specific questions, often using birds as models to test hypotheses or predictions based on theories. Most modern biological theories apply across taxonomic groups and the number of professional scientists who identify themselves as "ornithologists" has therefore declined. A wide range of tools and techniques are used in ornithology, both inside the laboratory and out in the field, and innovations are constantly made.

Etymology

The origins of the word *ornithology* come from the Greek *ornithologos* and late 17th-century Latin *ornithologia* meaning "bird science".

History

The history of ornithology largely reflects the trends in the history of biology, as well as many other scientific disciplines, including ecology, anatomy, physiology, paleontology, and more recently molecular biology. Trends include the move from mere descriptions to the identification of patterns, and thus towards elucidating the processes that produce these patterns.

Early Knowledge and Study

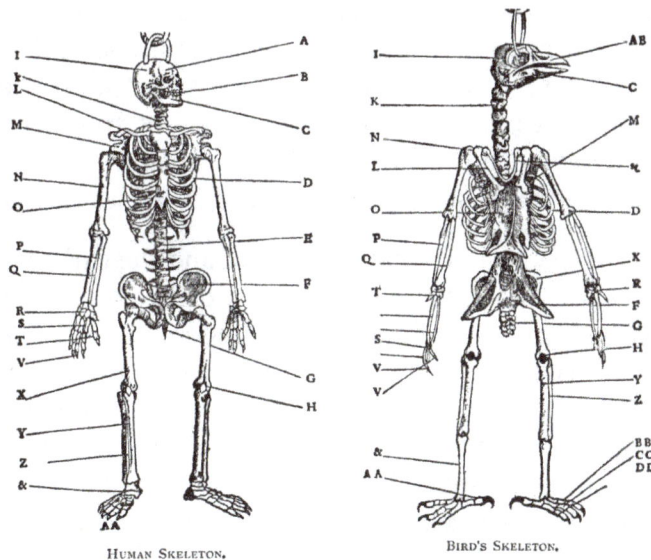

Belon's comparison of birds and humans in his *Book of Birds*, 1555

Humans have had an observational relationship with birds since prehistory, with some stone age drawings being amongst the oldest indications of an interest in birds. Birds were perhaps important as a food source, and bones of as many as 80 species have been found in excavations of early Stone Age settlements. Waterbird and seabird remains have also been found in shell mounds on the island of Oronsay off the coast of Scotland.

Cultures around the world have rich vocabularies related to birds. Traditional bird names are often based on detailed knowledge of the behaviour, with many names being onomatopoeic, many still in use. Traditional knowledge may also involve the use of birds in folk medicine and knowledge of these practices are passed on through oral traditions. Hunting of wild birds as well as their domestication would have required considerable knowledge of their habits. Poultry farming and falconry were practised from early times in many parts of the world. Artificial incubation of poultry was practised in China around 246 BC and around at least 400 BC in Egypt. The Egyptians also made use of birds in their hieroglyphic scripts, many of which, though stylized, are still identifiable to species.

Early written records provide valuable information on the past distributions of species. For instance Xenophon records the abundance of the ostrich in Assyria (Anabasis, i. 5); this subspecies from Asia minor is extinct and all extant ostrich races are today restricted to Africa. Other old writings such as the Vedas (1500–800 BC) demonstrate the careful observation of avian life histories and includes the earliest reference to the habit of brood parasitism by the Asian koel (*Eudynamys scolopacea*). Like writing, the early art of China, Japan, Persia and India also demonstrate knowledge, with examples of scientifically accurate bird illustrations.

Aristotle in 350 BC in his *Historia Animalium* noted the habit of bird migration, moulting, egg laying and life spans, as well as compiling a list of 170 different bird species. However, he also introduced and propagated several myths, such as the idea that swallows hibernated in winter, although he noted that cranes migrated from the steppes of Scythia to the marshes at the headwaters of the Nile. The idea of swallow hibernation became so well established that, even as late as in 1878, Elliott Coues could list as many as 182 contemporary publications dealing with the hibernation of swallows and little published evidence to contradict the theory. Similar misconceptions existed regarding the breeding of barnacle geese. Their nests had not been seen and it was believed that they grew by transformations of goose barnacles, an idea that became prevalent from around the 11th century and noted by Bishop Giraldus Cambrensis (Gerald of Wales) in *Topographia Hiberniae* (1187). Around 77 AD, Pliny the Elder described birds, among other creatures, in his *Historia Naturalis*.

The origins of falconry have been traced to Mesopotamia and the earliest record comes from the reign of Sargon II (722–705 BC). Falconry made its entry to Europe only after AD 400, brought in from the East after invasions by the Huns and Allans. Frederick II of Hohenstaufen (1194–1250) learned about Arabian falconry during wars in the region and obtained an Arabic treatise on falconry by Moamyn. He had this work translated into Latin and also conducted experiments on birds in his menagerie. By sealing the eyes of vultures and placing food nearby, he concluded that they found food by sight, and not by smell. He also developed methods to keep and train falcons. The studies that he undertook over nearly 30 years, were published in 1240 as De Arte Venandi cum Avibus (The Art of Hunting with Birds), considered one of the earliest studies on bird behaviour, and the first work known to include illustrations of birds.

Several early German and French scholars compiled old works and conducted new research on birds. These included Guillaume Rondelet who described his observations in the Mediterranean and Pierre Belon who described the fish and birds that he had seen in France and the Levant. Belon's Book of Birds (1555) is a folio volume with descriptions of some two hundred species. His comparison of the skeleton of humans and birds is considered as a landmark in comparative anatomy. Volcher Coiter (1534–1576), a Dutch anatomist made detailed studies of the internal structures of birds and produced a classification of birds, *De Differentiis Avium* (around 1572), that was based on structure and habits. Konrad Gesner wrote the *Vogelbuch* and *Icones avium omnium* around 1557. Like Gesner, Ulisse Aldrovandi, an encyclopedic naturalist began a 14-volume natural history with three volumes on birds, entitled *ornithologiae hoc est de avibus historiae libri XII* which was published from 1599 to 1603. Aldrovandi showed great interest in plants and animals and his work included 3000 drawings of fruits, flowers, plants and animals, published in 363 volumes. His *Ornithology* alone covers 2000 pages and included such aspects as the chicken and poultry techniques. He used a number of traits including behaviour, particularly bathing and dusting, to classify bird groups.

Cover of Ulisse Aldrovandi's *Ornithology*, 1599

Antonio Valli da Todi, who wrote on aviculture in 1601, knew the connections between territory and song

William Turner's *Historia Avium* ("History of Birds"), published at Cologne in 1544, was an early ornithological work from England. He noted the commonness of kites in English cities where they snatched food out of the hands of children. He included folk beliefs such as those of anglers. Anglers believed that the osprey emptied their fishponds and would kill them, mixing the flesh of the osprey into their fish bait. Turner's work reflected the violent times that he lived in and stands in contrast to later works such as Gilbert White's *The Natural History and Antiquities of Selborne* that were written in a tranquil era.

In the 17th century Francis Willughby (1635–1672) and John Ray (1627–1705) came up with the first major system of bird classification that was based on function and morphology rather than on form or behaviour. Willughby's *Ornithologiae libri tres* (1676) completed by John Ray is sometimes considered to mark the beginning of scientific ornithology. Ray also worked on *Ornithologia* which was published posthumously in 1713 as *Synopsis methodica avium et piscium*. The earliest list of British birds, *Pinax Rerum Naturalium Britannicarum* was written by Christopher Merrett in 1667, but authors such as John Ray considered it of little value. Ray did however, value the expertise of the naturalist Sir Thomas Browne (1605–82) who, not only answered his queries on ornithological identification and nomenclature, but also those of Willoughby and Merrett in letter correspondence. Browne himself in his lifetime kept an eagle, owl, cormorant, bittern and ostrich, penned a tract on falconry, and introduced the words *incubation* and *oviparous* into the English language.

An Experiment on a Bird in the Air Pump, Joseph Wright of Derby, 1768

Towards the late 18th century, Mathurin Jacques Brisson (1723–1806) and Comte de Buffon (1707–1788) began new works on birds. Brisson produced a six-volume work *Ornithologie* in 1760 and Buffon's included nine volumes (volumes 16–24) on birds *Histoire naturelle des oiseaux* (1770–1785) in his work on science *Histoire naturelle générale et particulière* (1749–1804). Coenraad Jacob Temminck (1778–1858) sponsored François Le Vaillant [1753–1824] to collect bird specimens in Africa and this resulted in Le Vaillant's six-volume *Histoire naturelle des oiseaux d'Afrique* (1796–1808). Louis Jean Pierre Vieillot (1748–1831) spent ten years studying North American birds and wrote the *Histoire naturelle des oiseaux de l'Amerique septentrionale* (1807–1808?). Vieillot pioneered in the use of life-histories and habits in classification. Alexander Wilson composed a nine-volume work, American Ornithology, published 1808-14—the first such record of North American birds, significantly predating Audubon. In the early 19th century, Lewis and Clark studied and identified many birds in the western United States. John James Audubon, born in 1785, observed and painted birds in France and later in the Ohio and Mississippi valleys. From 1827 to 1838, Audubon published *The Birds of America*, which was engraved by Robert Havell, Sr. and his son Robert Havell, Jr.. Containing 435 engravings, it is often regarded as the greatest ornithological work in history.

Scientific Studies

The emergence of ornithology as a scientific discipline began in the 18th century when Mark Catesby published his two-volume *Natural History of Carolina, Florida and the Bahama Islands*, a landmark work which included 220 hand-painted engravings and was the basis for many of the species Carl Linnaeus described in the 1758 *Systema Naturae*. Linnaeus' work revolutionised bird taxonomy by assigning every species a binomial name, categorising them into different genera. However, it was not until the Victorian era—with the concept of natural history, and the collection of natural objects such as bird eggs and skins—that ornithology emerged as a specialised science. This specialization led to the formation in Britain of the British Ornithologists' Union in 1858. In 1859 the members founded its journal *The Ibis*. The sudden spurt in ornithology was also due in part to colonialism. A hundred years later, in 1959, R. E. Moreau noted that ornithology in this period was preoccupied with the geographical distributions of various species of birds.

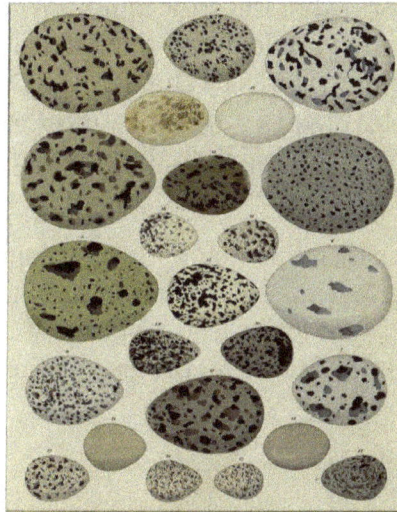

Early bird study focused on collectibles such as eggs and nests

No doubt the preoccupation with widely extended geographical ornithology, was fostered by the immensity of the areas over which British rule or influence stretched during the 19th century and for some time afterwards.

—Moreau

The bird collectors of the Victorian era observed the variations in bird forms and habits across geographic regions, noting local specialization and variation in widespread species. The collections of museums and private collectors grew with contributions from various parts of the world. The naming of species with binomials and the organization of birds into groups based on their similarities became the main work of museum specialists. The variations in widespread birds across geographical region caused the introduction of trinomial names.

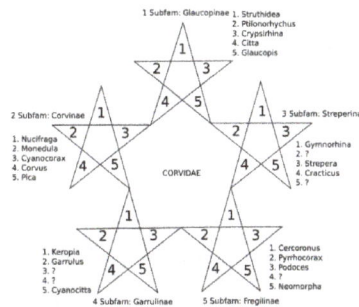

Kaup's classification of the crow family

The search for patterns in the variations of birds was attempted by many. Friedrich Wilhelm Joseph Schelling (1775–1854), his student Johann Baptist von Spix (1781–1826) and several others believed that there was a hidden and innate mathematical order in the forms of birds. They believed that there was a "natural" classification that was superior to "artificial" ones. A particularly popular idea was the Quinarian system popularised by Nicholas Aylward Vigors (1785–1840), William Sharp Macleay (1792–1865), William Swainson and others. The idea was that nature followed a "rule of five" with five groups nested hierarchically. Some had attempted a rule of four, but Johann Jakob Kaup (1803–1873) insisted that the number five was special noting that other natural entities such as the senses also came in fives. He followed this idea and demonstrated his view of the order within the crow family. Where he failed to find 5 genera, he left a blank insisting that a new genus would found to fill these gaps. These ideas were replaced by more complex "maps" of affinities in works by Hugh Edwin Strickland and Alfred Russel Wallace. A major advance was made by Max Fürbringer in 1888 who established a comprehensive phylogeny of birds based on anatomy, morphology, distribution and biology. This was developed further by Hans Gadow and others.

The Galapagos finches were especially influential in the development of Charles Darwin's theory of evolution. His contemporary Alfred Russel Wallace also noted these variations and the geographical separations between different forms leading to the study of biogeography. Wallace was influenced by the work of Philip Lutley Sclater on the distribution patterns of birds.

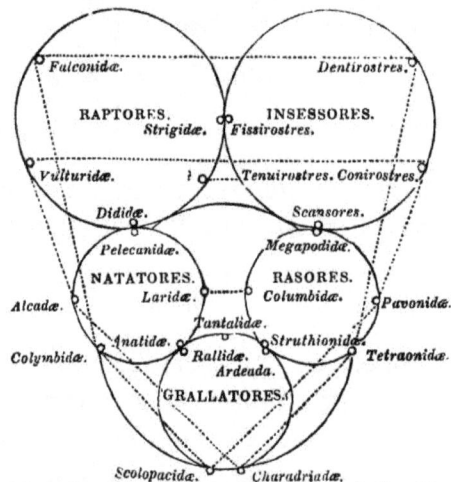

Affinities and analogies among the groups according to Swainson. The circles touch with groups on them having "affinities", but the lines connect groups that showed "analogies".

Quinarian system of bird classification by Swainson

For Darwin, the problem was how species arose from a common ancestor, but he did not attempt to find rules for delineation of species. The species problem was tackled by the ornithologist Ernst Mayr. Mayr was able to demonstrate that geographical isolation and the accumulation of genetic differences led to the splitting of species.

Early ornithologists were preoccupied with matters of species identification. Only systematics counted as true science and field studies were considered inferior through much of the 19th century. In 1901 Robert Ridgway wrote in the introduction to *The Birds of North and Middle America* that:

There are two essentially different kinds of ornithology: systematic or scientific, and popular. The former deals with the structure and classification of birds, their synonymies and technical descriptions. The latter treats of their habits, songs, nesting, and other facts pertaining to their life histories.

This early idea that the study of *living birds* was merely recreation held sway until ecological theories became the predominant focus of ornithological studies. The study of birds in their habitats was particularly advanced in Germany with bird ringing stations established as early as 1903. By the 1920s the *Journal für Ornithologie* included many papers on the behaviour, ecology, anatomy and physiology, many written by Erwin Stresemann. Stresemann changed the editorial policy of the journal, leading both to a unification of field and laboratory studies and a shift of research from museums to universities. Ornithology in the United States continued to be dominated by museum studies of morphological variations, species identities and geographic distributions, until it was influenced by Stresemann's student Ernst Mayr. In Britain, some of the earliest ornithological works that used the word ecology appeared in 1915. *The Ibis* however resisted the introduction of these new methods of study and it was not until 1943 that any paper on ecology appeared. The work of David Lack on population ecology was pioneering. Newer quantitative approaches were introduced for the study of ecology and behaviour and this was not readily accepted. For instance, Claud Ticehurst wrote:

Sometimes it seems that elaborate plans and statistics are made to prove what is commonplace knowledge to the mere collector, such as that hunting parties often travel more or less in circles.

— Ticehurst

David Lack's studies on population ecology sought to find the processes involved in the regulation of population based on the evolution of optimal clutch sizes. He concluded that population was regulated primarily by density-dependent controls, and also suggested that natural selection produces life-history traits that maximize the fitness of individuals. Others like Wynne-Edwards interpreted population regulation as a mechanism that aided the "species" rather than individuals. This led to widespread and sometimes bitter debate on what constituted the "unit of selection". Lack also pioneered the use of many new tools for ornithological research, including the idea of using radar to study bird migration.

Birds were also widely used in studies of the niche hypothesis and Georgii Gause's competitive exclusion principle. Work on resource partitioning and the structuring of bird communities through competition were made by Robert MacArthur. Patterns of biodiversity also became a topic of interest. Work on the relationship of the number of species to area and its application in the study of island biogeography was pioneered by E. O. Wilson and Robert MacArthur. These studies led to the development of the disciplin e of landscape ecology.

John Hurrell Crook studied the behaviour of weaverbirds and demonstrated the links between ecological conditions, behaviour and social systems. Principles from economics were introduced to the study of biology by Jerram L. Brown in his work on explaining territorial behaviour. This led to more studies of behaviour that made use of cost-benefit analyses. The rising interest in sociobiology also led to a spurt of bird studies in this area.

A mounted specimen of a red-footed falcon.

The study of imprinting behaviour in ducks and geese by Konrad Lorenz and the studies of instinct in herring gulls by Nicolaas Tinbergen, led to the establishment of the field of ethology. The study of learning became an area of interest and the study of bird song has been a model for studies in neuro-ethology. The study of hormones and physiology in the control of behaviour has also been aided by bird models. These have helped in finding the proximate causes of circadian and seasonal cycles. Studies on migration have attempted to answer questions on the evolution of migration, orientation and navigation.

The growth of genetics and the rise of molecular biology led to the application of the gene-centered view of evolution to explain avian phenomena. Studies on kinship and altruism, such as helpers, became of particular interest. The idea of inclusive fitness was used to interpret observations on behaviour and life-history and birds were widely used models for testing hypotheses based on theories postulated by W. D. Hamilton and others.

The new tools of molecular-biology changed the study of bird systematics. Systematics changed from being based on phenotype to the underlying genotype. The use of techniques such as DNA-DNA hybridization to study evolutionary relationships was pioneered by Charles Sibley and Jon Edward Ahlquist resulting in what is called the Sibley-Ahlquist taxonomy. These early techniques have been replaced by newer ones based on mitochondrial DNA sequences and molecular phylogenetics approaches that make use of computational procedures for sequence alignment, construction of phylogenetic trees and calibration of molecular clocks to infer evolutionary relationships. Molecular techniques are also widely used in studies of avian population biology and ecology.

Rise to Popularity

The use of field glasses or telescopes for bird observation began in the 1820s and 1830s with pioneers like J. Dovaston (who also pioneered in the use of bird-feeders), but it was not until the

1880s that instruction manuals began to insist on the use of optical aids such as "a first-class tele-scope" or "field glass."

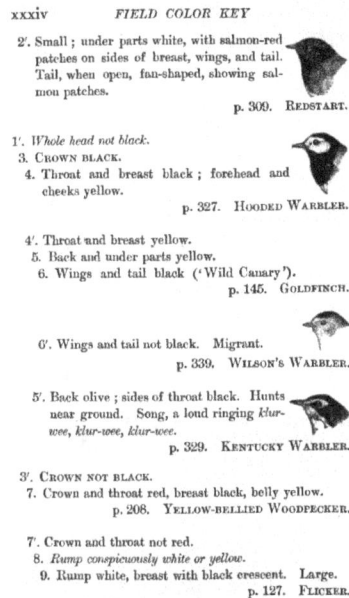

xxxiv   *FIELD COLOR KEY*

2'. Small ; under parts white, with salmon-red patches on sides of breast, wings, and tail. Tail, when open, fan-shaped, showing salmon patches.
p. 309. REDSTART.

1'. *Whole head not black.*
3. CROWN BLACK.
4. Throat and breast black ; forehead and cheeks yellow.
p. 327. HOODED WARBLER.

4'. Throat and breast yellow.
5. Back and under parts yellow.
6. Wings and tail black ('Wild Canary').
p. 145. GOLDFINCH.

6'. Wings and tail not black. Migrant.
p. 339. WILSON'S WARBLER.

5'. Back olive ; sides of throat black. Hunts near ground. Song, a loud ringing *klur-wee, klur-wee, klur-wee.*
p. 329. KENTUCKY WARBLER.

3'. CROWN NOT BLACK.
7. Crown and throat red, breast black, belly yellow.
p. 208. YELLOW-BELLIED WOODPECKER.

7'. Crown and throat not red.
8. *Rump conspicuously white or yellow.*
9. Rump white, breast with black crescent. Large.
p. 127. FLICKER.

Page from an early field guide by Florence Augusta Merriam Bailey

The rise of field guides for the identification of birds was another major innovation. The early guides such as those of Thomas Bewick (2 volumes) and William Yarrell (3 volumes) were cumbersome, and mainly focused on identifying specimens in the hand. The earliest of the new generation of field guides was prepared by Florence Merriam, sister of Clinton Hart Merriam, the mammalogist. This was published in 1887 in a series *Hints to Audubon Workers:Fifty Birds and How to Know Them* in Grinnell's *Audubon Magazine*. These were followed by new field guides including classics by Roger Tory Peterson.

The interest in birdwatching grew in popularity in many parts of the world and it was realized that there was a possibility for amateurs to contribute to biological studies. As early as 1916, Julian Huxley wrote a two part article in *The Auk*, noting the tensions between amateurs and professionals and suggested the possibility that the "vast army of bird-lovers and bird-watchers could begin providing the data scientists needed to address the fundamental problems of biology."

Organizations were started in many countries and these grew rapidly in membership, most notable among them being the Royal Society for the Protection of Birds (RSPB) in Britain and the Audubon Society in the US. The Audubon Society started in 1885. Both these organizations were started with the primary objective of conservation. The RSPB, born in 1889, grew from a small group of women in Croydon who met regularly and called themselves the "Fur, Fin and Feather Folk" and who took a pledge "to refrain from wearing the feathers of any birds not killed for the purpose of food, the Ostrich only exempted." The organization did not allow men as members initially, avenging a policy of the British Ornithologists' Union to keep out women. Unlike the RSPB, which was primarily conservation oriented, the British Trust for Ornithology (BTO) was started in 1933 with the aim of advancing ornithological research. Members were often involved in collaborative ornithological projects. These projects have resulted in atlases which detail the distribution of bird species across Britain. In the United States, the Breeding Bird Surveys, conducted by the US Geological Survey have also produced atlases with information on breeding densities and changes in the density and distribution over time. Other volunteer collaborative ornithology projects were subsequently established in other parts of the world.

Techniques

The tools and techniques of ornithology are varied and new inventions and approaches are quickly incorporated. The techniques may be broadly dealt under the categories of those that are applicable to specimens and those that are used in the field, however the classification is rough and many analysis techniques are usable both in the laboratory and field or may require a combination of field and laboratory techniques.

Collections

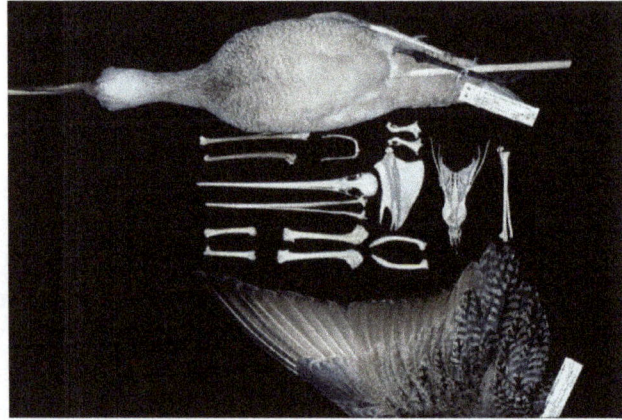

Bird preservation techniques

The earliest approaches to modern bird study involved the collection of eggs, a practice known as oology. While collecting became a pastime for many amateurs, the labels associated with these early egg collections made them unreliable for the serious study of bird breeding. In order to preserve eggs, a tiny hole was pierced and the contents extracted. This technique became standard with the invention of the blow drill around 1830. Egg collection is no longer popular; however historic museum collections have been of value in determining the effects of pesticides such as DDT on physiology. Museum bird collections continue to act as a resource for taxonomic studies.

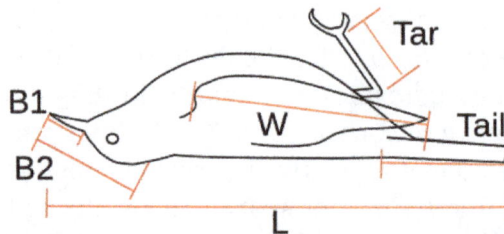

Morphometric measurements of birds are important in systematics

The use of bird skins to document species has been a standard part of systematic ornithology. Bird skins are prepared by retaining the key bones of the wings, leg and skull along with the skin and feathers. In the past, they were treated with arsenic to prevent fungal and insect (mostly dermestid) attack. Arsenic, being toxic, was replaced by borax. Amateur and professional collectors became familiar with these skinning techniques and started sending in their skins to museums, some of them from distant locations. This led to the formation of huge collections of bird skins in museums in Europe and North America. Many private collections were also formed. These became references for comparison of species and the ornithologists at these museums were able to compare species from different locations, often places that they themselves never visited. Morphometrics of these skins, particularly the lengths of the tarsus, bill, tail and wing became important in the descriptions of bird species. These skin collections have been utilized in more recent times for studies on molecular phylogenetics by the extraction of ancient DNA. The importance of type specimens in the description of species make skin collections a vital resource for systematic ornithology. However, with the rise of molecular techniques, it has now become possible to establish the taxonomic status of new discoveries, such as the Bulo Burti boubou (*Laniarius liberatus*, no longer a valid species) and the Bugun liocichla (*Liocichla bugunorum*), using blood, DNA and feather samples as the holotype material.

Other methods of preservation include the storage of specimens in spirit. Such wet-specimens have special value in physiological and anatomical study, apart from providing better quality of DNA for molecular studies. Freeze drying of specimens is another technique that has the advantage of preserving stomach contents and anatomy, although it tends to shrink making it less reliable for morphometrics.

In the Field

The study of birds in the field was helped enormously by improvements in optics. Photography made it possible to document birds in the field with great accuracy. High power spotting scopes today allow observers to detect minute morphological differences that were earlier possible only by examination of the specimen *in the hand*.

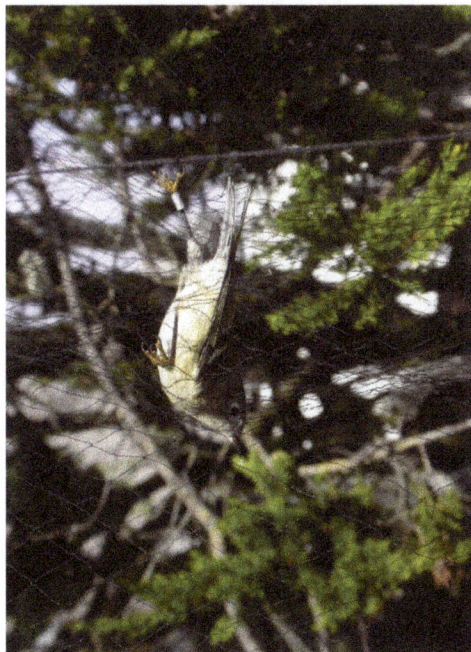

A bird caught in a mist net

The capture and marking of birds enables detailed studies of life-history. Techniques for capturing birds are varied and include the use of bird liming for perching birds, mist nets for woodland birds, cannon netting for open area flocking birds, the bal-chatri trap for raptors, decoys and funnel traps for water birds.

A California condor marked with wing tags

The bird in the hand may be examined and measurements can be made including standard lengths and weight. Feather moult and skull ossification provide indications of age and health. Sex can be determined by examination of anatomy in some sexually non-dimorphic species. Blood samples may be drawn to determine hormonal conditions in studies of physiology, identify DNA markers for studying genetics and kinship in studies of breeding biology and phylogeography. Blood may also be used to pathogens and arthropod borne viruses. Ectoparasites may be collected for studies of coevolution and zoonoses. In many of cryptic species, measurements (such as the relative lengths of wing feathers in warblers) are vital in establishing identity.

Captured birds are often marked for future recognition. Rings or bands provide long-lasting identification but require capture for the information on them to be read. Field identifiable marks such as coloured bands, wing tags or dyes enable short-term studies where individual identification is required. Mark and recapture techniques make demographic studies possible. Ringing has traditionally been used in the study of migration. In recent times satellite transmitters provide the ability to track migrating birds in near real-time.

Techniques for estimating population density include point counts, transects and territory mapping. Observations are made in the field using carefully designed protocols and the data may be analysed to estimate bird diversity, relative abundance or absolute population densities. These methods may be used repeatedly over large time spans to monitor changes in the environment. Camera traps have been found to be a useful tool for the detection and documentation of elusive species, nest predators and in the quantitative analysis of frugivory, seed dispersal and behaviour.

In the Laboratory

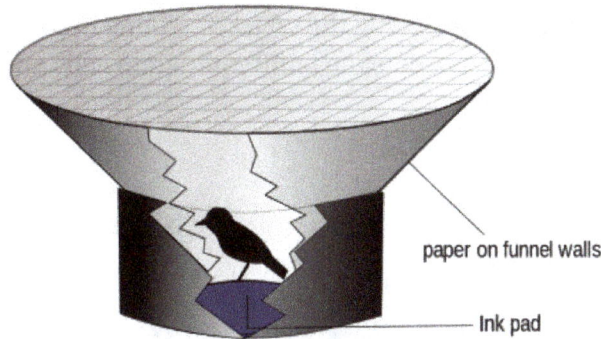

An Emlen funnel is used to study the orientation behaviour of migratory birds in a laboratory. Experimenters sometimes place the funnel inside a planetarium to study night migration.

Many aspects of bird biology are difficult to study in the field. These include the study of behavioural and physiological changes that require a long duration of access to the bird. Non-destructive samples of blood or feathers taken during field studies may be studied in the laboratory. For instance, the variation in the ratios of stable hydrogen isotopes across latitudes makes it possible to roughly establish the origins of migrant birds using mass spectrometric analysis of feather samples. These techniques can be used in combination with other techniques such as ringing.

The first attenuated vaccine developed by Louis Pasteur was for fowl cholera and was tested on poultry in 1878. Poultry continues to be used as a model for many studies in non-mammalian immunology.

Studies in bird behaviour include the use of tamed and trained birds in captivity. Studies on bird intelligence and song learning have been largely laboratory based. Field researchers may make use of a wide range of techniques such as the use of dummy owls to elicit mobbing behaviour, dummy males or the use of call playback to elicit territorial behaviour and thereby to establish the boundaries of bird territories.

Studies of bird migration including aspects of navigation, orientation and physiology are often studied using captive birds in special cages that record their activities. The Emlen funnel for instance makes use of a cage with an inkpad at the centre and a conical floor where the ink marks can be counted to identify the direction in which the bird attempts to fly. The funnel can have a transparent top and visible cues such as the direction of sunlight may be controlled using mirrors or the positions of the stars simulated in a planetarium.

The entire genome of the domestic fowl (*Gallus gallus*) was sequenced in 2004 and was followed in 2008 by the genome of the zebra finch (*Taeniopygia guttata*). Such whole genome sequencing projects allow for studies on evolutionary processes involved in speciation. Associations between the expression of genes and behaviour may be studied using candidate genes. Variations in the exploratory behaviour of great tits (*Parus major*) have been found to be linked with a gene orthologous to the human gene *DRD4* (Dopamine receptor D4) which is known to be associated with novelty-seeking behaviour. The role of gene expression in developmental differences and morphological variations have been studied in Darwin's finches. The difference in the expression of *Bmp4* have been shown to be associated with changes in the growth and shape of the beak.

The chicken has long been a model organism for studying vertebrate developmental biology. As the embryo is readily accessible, its development can be easily followed (unlike mice). This also allows the use of electroporation for studying the effect of adding or silencing a gene. Other tools for perturbing their genetic makeup are chicken embryonic stem cells and viral vectors.

Collaborative Studies

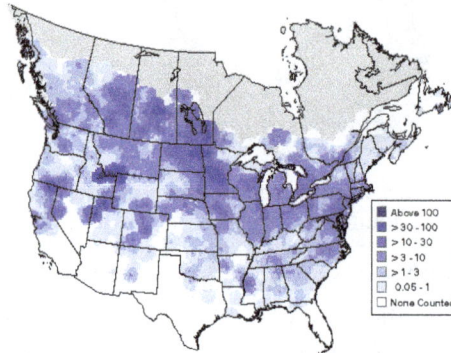

Summer distribution and abundance of Canada goose using data from the North American Breeding Bird Surveys 1994–2003

With the widespread interest in birds, it has been possible to use a large number of people to work on collaborative ornithological projects that cover large geographic scales. These citizen science projects include nationwide projects such as the Christmas Bird Count, Backyard Bird Count, the North American Breeding Bird Survey, the Canadian EPOQ or regional projects such as the Asian Waterfowl Census and Spring Alive in Europe. These projects help to identify distributions of birds, their population densities and changes over time, arrival and departure dates of migration, breeding seasonality and even population genetics. The results of many of these projects are published as bird atlases. Studies of migration using bird ringing or colour marking often involve the cooperation of people and organizations in different countries.

Applications

Wild birds impact many human activities while domesticated birds are important sources of eggs, meat, feathers and other products. Applied and economic ornithology aim to reduce the ill effects of problem birds and enhance gains from beneficial species.

Red-billed quelea are a major agricultural pest in parts of Africa.

The role of some species of birds as pests has been well known, particularly in agriculture. Granivorous birds such as the queleas in Africa are among the most numerous birds in the world and foraging flocks can cause devastation. Many insectivorous birds are also noted as beneficial in agriculture. Many early studies on the benefits or damages caused by birds in fields were made by analysis of stomach contents and observation of feeding behaviour. Modern studies aimed to manage birds in agriculture make use of a wide range of principles from ecology. Intensive aquaculture has brought humans in conflict with fish-eating birds such as cormorants.

Large flocks of pigeons and starlings in cities are often considered as a nuisance and techniques to reduce their populations or their impacts are constantly innovated. Birds are also of medical importance and their role as carriers of human diseases such as Japanese Encephalitis, West Nile Virus and H5N1 have been widely recognised. Bird strikes and the damage they cause in aviation are of particularly great importance, due to the fatal consequences and the level of economic losses caused. It has been estimated that the airline industry incurs worldwide damages of US$1.2 billion each year.

Many species of birds have been driven to extinction by human activities. Being conspicuous elements of the ecosystem, they have been considered as indicators of ecological health. They have also helped in gathering support for habitat conservation. Bird conservation requires specialized knowledge in aspects of biology, ecology and may require the use of very location specific approaches. Ornithologists contribute to conservation biology by studying the ecology of birds in the wild and identifying the key threats and ways of enhancing the survival of species. Critically endangered species such as the California condor have had to be captured and bred in captivity. Such ex-situ conservation measures may be followed by re-introduction of the species into the wild.

Paleozoology

Priscacara liops

Palaeozoology, also spelled as Paleozoology, is the branch of paleontology, paleobiology, or zoology dealing with the recovery and identification of multicellular animal remains from geological (or even archeological) contexts, and the use of these fossils in the reconstruction of prehistoric environments and ancient ecosystems.

Definitive, macroscopic remains of these metazoans are found in the fossil record from the Ediacaran period of the Neoproterozoic era onwards, although they do not become common until the Late Devonian period in the latter half of the Paleozoic era.

Perhaps the best known macrofossils group is the dinosaurs. Other popularly known animal-derived macrofossils include trilobites, crustaceans, echinoderms, brachiopods, mollusks, bony fishes, sharks, Vertebrate teeth, and shells of numerous invertebrate groups. This is because hard organic parts, such as bones, teeth, and shells resist decay, and are the most commonly preserved and found animal fossils. Exclusively soft-bodied animals—such as jellyfish, flatworms, nematodes, and insects—are consequently rarely fossilized, as these groups do not produce hard organic parts.

Vertebrate Paleozoology

Vertebrate paleozoology refers to the use of morphological, temporal, and stratigraphic data to map vertebrate history in evolutionary theory. Vertebrates are classified as a subphylum of Chordata, a phylum used to classify species adhering to a rod-shaped, flexible body type called a notochord. They differ from other phyla in that other phyla may have cartilage or cartilage-like tissues forming a sort of skeleton, but only vertebrates possess what we define as bone.

Classes of vertebrates listed in chronological order from oldest to most recent include heterostracans, osteostracans, coelolepid agnathans, acanthodians, osteichthyan fishes, chondrichthyan fishes, amphibians, reptiles, mammals, and birds. All vertebrates are studied under standard evolutionary generalizations of behavior and life process, although there is controversy over whether population can be accurately estimated from limited fossil resources.

Evolutionary origins of vertebrates as well as the phylum Chordata have not been scientifically determined. Many believe vertebrates diverged from a common ancestor of chordates and echinoderms. This belief is well supported by the prehistoric marine creature Amphioxus. Amphioxus does not possess bone, making it an invertebrate, but it has common features with vertebrates including a segmented body and a notochord. This could imply that Amphioxus is a transitional form between an early chordate, echinoderm or common ancestor, and vertebrates.

Quantitative Paleozoology

Quantitative paleozoology is a process of taking a census of fossil types rather than inventory. They differ in that inventory refers to a detailed log of individual fossils, whereas census attempts to group individual fossils to tally the total number of a species. This information can be used to determine which species were most dominant and which had the largest population at a time period or in a geological region.

In the early 1930s, paleontologists Chester Stock and Hildegarde Howard devised special units for quantitative paleozoology and quantitative paleontology. The first unit used, Number of Identified

Species (NISP), specified exact quantity of fossils from a specific species recorded. Stock and How-ard determined this unit to be problematic for quantitative purposes as an excess of a small fossil—such as teeth—could exaggerate quantity of the species. There was also an amount of confusion as to whether bone fragments should be assembled and counted as one bone or tallied individually. Stock and Howard then devised the Minimum Number of Individuals (MNI), which estimated the minimum number of animals needed to produce the fossils recorded. For example, if five scapulae from a species were found, it might be difficult to determine whether some of them were paired right and left on one individual or whether each came from a different individual, which could alter census, but it could be said that there must be at least three individuals to produce five scapulae. Five would thus be the MNI. In rare cases where enough of a collection of fossils can be assembled into individuals as to provide an accurate number of individuals, the unit used is Actual Number of Individuals, or ANI.

Another unit commonly used in quantitative paleozoology is biomass. Biomass is defined as the amount of tissue in an area or from a species. It is calculated by estimating an average weight based on similar modern species and multiplying it by the MNI. This yields an estimate of how much the entire population of a species may have weighed. Problems with this measurement in-clude the difference in weight between youngsters and adults, seasonal weight changes due to diet and hibernation, and the difficulty of accurately estimating the weight of a creature with only a skeletal reference. It is also difficult to determine exact age of fossilized matter within a year or a decade, so a biomass might be grossly exaggerated or under exaggerated if the estimated time frame in which the fossils were alive is incorrect.

A similar measurement to biomass is meat weight. To determine meat weight, MNI is multiplied by the amount of meat an individual is thought to have provided, then multiplied by the percent-age of that meat thought to be edible. This gives an estimate of "pounds of usable meat" per indi-vidual which might have been harvested by prehistoric hunters. For example, a male Wapiti has an average weight of 400 kg, and in a particular study, the MNI of Wapiti was found to be 10. This would create a biomass of 4,000 kg. If the amount of edible meat is estimated at 50 percent, this would result in a meat weight of 2,000 kg. The biggest problem with this method is the debate over "percent of usable meat." Different views on which parts of a species are edible and which are not as well as whether or not primitive butchers would have been able to access and prepare different parts have led to controversy.

Conservation Biology

Paleozoological data is used in research concerning conservation biology. Conservation biology refers to biological study used for conservation, control, and preservation of various species and ecosystems. In this context, the paleozoological data used is obtained from recently deceased de-composing matter rather than prehistoric matter.

R. Lee Lyman, Professor and Chair Department of Anthropology at the University of Missouri, writes that paleozoological research can provide data such as extinction rates and causes and "benchmark" peaks and drops in population which can be used to predict future patterns and to design maximally effective methods of controlling these patterns. In addition, paleozoological data can be used to compare current to former population and distribution of a species.

Cetology

A researcher fires a biopsy dart at an orca. The dart will remove a small piece of the whale's skin and bounce harmlessly off the animal.

Cetology or Whalelore is the branch of marine mammal science that studies the approximately eighty species of whales, dolphins, and porpoise in the scientific order Cetacea. Cetologists, or those who practice cetology, seek to understand and explain cetacean evolution, distribution, morphology, behavior, community dynamics, and other topics.

History

Bottlenose dolphin

Observations about Cetacea have been recorded since at least classical times. Ancient Greek fishermen created an artificial notch on the dorsal fin of dolphins entangled in nets so that they could tell them apart years later.

Approximately 2,300 years ago, Aristotle carefully took notes on cetaceans while traveling on boats with fishermen in the Aegean Sea. In his book *Historia animalium* (History of animals), Aristotle was careful enough to distinguish between the baleen whales and toothed whales, a taxonomical separation still used today. He also described the Sperm Whale and the common dolphin, stating that they can live for at least twenty-five or thirty years. His achievement was remarkable for its time, because even today it is very difficult to estimate the life-span of advanced marine animals. After Aristotle's death, much of the knowledge he had gained about cetaceans was lost, only to be re-discovered during the Renaissance.

Many of the medieval texts on cetaceans come mainly from Scandinavia and Iceland, most came about the mid-13th century. One of the better known is *Speculum Regale*. In this text is described various species that lived around the island of Iceland. It mentions "orcs" that had dog-like teeth and would demonstrate the same kind of aggression towards other cetaceans as wild dogs would to other terrestrial animals. The text even illustrated the hunting technique of Orcs, which are now called Orcas. The *Speculum Regale* describes other cetaceans, including the Sperm Whale and Narwhal. Many times they were seen as terrible monsters, such as killers of men, and destroyers of ships. They even bore odd names such as "Pig Whale", "Horse Whale", and "Red Whale". But not all creatures described were said to be fierce. Some were seen to be good, such as whales that drove shoals of herring towards the shore. This was seen as very helpful to fisherman.

Many of the early studies were based on dead specimens and myth. The little information that was gathered was usually length, and a rough outer body anatomy. Because these animals live in water their entire lives, early scientists did not have the technology to go study these animals further. It was not until the 16th century that things would begin to change. Then cetaceans would be proved to be mammals rather than fish.

Aristotle argued they were mammals. But Pliny the Elder stated that they were fish, and it was followed by many naturalists. However, Pierre Belon (1517–1575) and G. Rondelet (1507–1566) persisted on convincing they were mammals. They argued that the animals had lungs and a uterus, just like mammals. Not until 1758, when Swedish botanist Carl Linnaeus (1707–1778) published the tenth edition of *Systema Naturae*, were they seen as mammals.

Only decades later, French zoologist and paleontologist Baron Georges Cuvier (1769–1832) described the animals as mammals without any hind legs. Skeletons were assembled and displayed in the first natural history museums, and on a closer look and comparisons with other extinct animal fossils led zoologists to conclude that cetaceans came from a family of ancient land mammals.

Between the 16th–20th century, much of our information on cetaceans came from whalers. Whalers were the most knowledgeable about the animals, but their information was regarding migration routes and outer anatomy, and only little information of behavior. During the 1960s, people began studying the animals intensively, often in dedicated research institutes. The Tethys Institute of Milan, founded in 1986, compiled an extensive cetology database of the Mediterranean. This came from both concern about wild populations and also the capture of larger animals such as the Orca, and gaining popularity of dolphin shows in marine parks.

Studying Cetaceans

Studying cetaceans presents numerous challenges. Cetaceans only spend 10% of their time on the surface, and all they do at the surface is breathe. There is very little behavior seen at the surface. It is also impossible to find any signs that an animal has been in an area. Cetaceans do not leave tracks that can be followed. However, the dung of whales often floats and can be collected to tell important information about their diet and about the role they have in the environment. Many times cetology consists of waiting and paying close attention.

Humpback Whales often have distinct markings that enable scientists to identify individuals.

Cetologists use equipment including hydrophones to listen to calls of communicating animals, binoculars and other optical devices for scanning the horizon, cameras, notes, and a few other devices and tools.

An alternative method of studying cetaceans is through examination of dead carcasses that wash up on the shore. If properly collected and stored, these carcasses can provide important information that is difficult to obtain in field studies.

Identifying Individuals

In recent decades, methods of identifying individual cetaceans have enabled accurate population counts and insights into the lifecycles and social structures of various species.

One such successful system is photo-identification. This system was popularized by Michael Bigg, a pioneer in modern orca (killer whale) research. During the mid-1970s, Bigg and Graeme Ellis photographed local orcas in the British Columbian seas. After examining the photos, they realized they could recognize certain individual whales by looking at the shape and condition of the dorsal fin, and also the shape of the saddle patch. These are as unique as a human fingerprint; no one animal's looks is exactly like another's. After they could recognize certain individuals, they found that the animals travel in stable groups called pods. Researchers use photo identification to identify specific individuals and pods.

The photographic system has also worked well in humpback whale studies. Researchers use the color of the pectoral fins and color of the fluke to identify individuals. Scars from orca attacks found on the flukes of humpbacks are also used in identification.

No branch of zoology is so much involved as that which is entitled Cetology.

— *William Scoresby (as quoted in Moby-Dick)*

Nematology

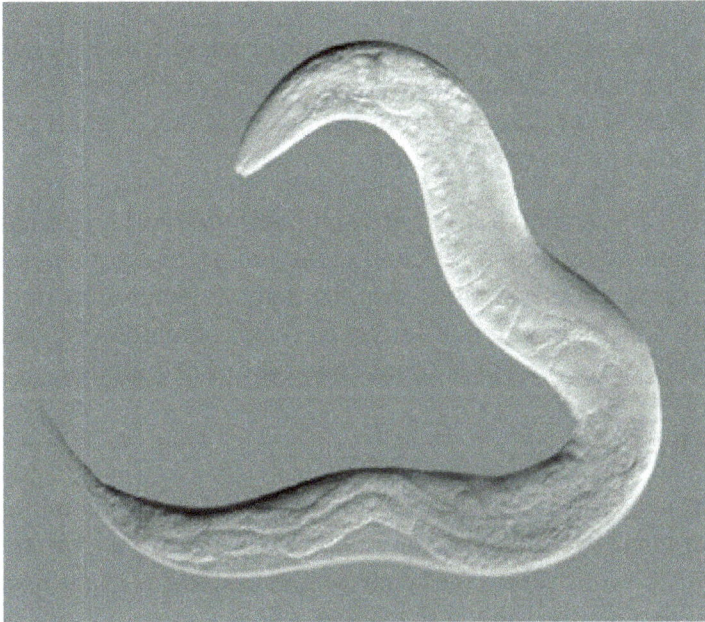

C. elegans

Nematology is the scientific discipline concerned with the study of nematodes, or roundworms. Although nematological investigation dates back to the days of Aristotle or even earlier, nematology as an independent discipline has its recognizable beginnings in the mid to late 19th century.

History: Pre-1850

Nematology research, like most fields of science, has its foundations in observations and the recording of these observations. The earliest written account of a nematode "sighting," as it were, may be found in the Pentateuch of the Old Testament in the Bible, in the Fourth Book of Moses called Numbers: "And the Lord sent fiery serpents among the people, and they bit the people; and much people of Israel died". Although no empirical data exists to test the hypothesis, many nematologists assume and circumstantial evidence suggests the "fiery serpents" to be the Guinea worm, *Dracunculus medinensis*, as this nematode is known to inhabit the region near the Red Sea.

Before 1750, a large number of nematode observations were recorded, many by the notable great minds of ancient civilization. Hippocrates (ca. 420 B.C.), Aristotle (ca. 350 B.C.), Celsus (ca 10 B.C.), Galen (ca. 180 A.D.) and Redi (1684) all described nematodes parasitizing humans or other large animals and birds. Borellus (1653) was the first to observe and describe a free-living nematode, which he dubbed the "vinegar eel;" and Tyson (1683) used a crude microscope to describe the rough anatomy of the human intestinal roundworm, *Ascaris lumbricoides*.

Other well-known microscopists spent time observing and describing free-living and animal-parasitic nematodes: Hooke (1683), Leeuwenhoek (1722), Needham (1743), and Spallanzani (1769) are among these. Observations and descriptions of plant parasitic nematodes, which were less conspicuous to ancient scientists, didn't receive as much or as early attention as did animal parasites. The earliest allusion to a plant parasitic nematode is, however, preserved in famous writ. "Sowed cockle, reap'd no corn," a line by William Shakespeare penned in 1594 in *Love's Labour's Lost*, Act IV, Scene 3, most certainly has reference to blighted wheat caused by the plant parasite, *Anguina tritici*.

Needham (1743) solved the "riddle of cockle" when he crushed one of the diseased wheat grains and observed "Aquatic Animals...denominated Worms, Eels, or Serpents, which they very much resemble." It is likely that few or no other recorded observations of plant parasitic nematodes or their effects are to be found in ancient literature.

From 1750 to the early 1900s, nematology research continued to be descriptive and taxonomic, focusing primarily on free-living nematodes and plant and animal parasites. During this period a number of productive researchers contributed to the field of nematology in the United States and abroad. Beginning with Needham and continuing to Cobb, nematologists compiled and continuously revised a broad descriptive morphological taxonomy of nematodes.

History: 1850 to the Present

Kuhn (1874) is thought to be the first to use soil fumigation to control nematodes, applying carbon disulfide treatments in sugar beet fields in Germany. In Europe from 1870 to 1910, nematological research focused heavily on controlling the sugar beet nematode as sugar beet production became an important economy during this time in the Old World.

Although 18th and 19th century scientists yielded a considerable amount of important fundamental and applied knowledge about nematode biology, nematology research really began to advance in quality and quantity near the turn of the 20th century. In 1918, the first permanent nematology field station was constructed in the U.S. Post Office in Salt Lake City, Utah under the direction of Harry B. Shaw, after scientists observed the sugar beet nematode in a field south of the city. In this same year, Nathan Cobb (1918) published his Contributions to a Science of Nematology and his lab manual "Estimating the Nema Population of Soil." These two publications provide definitive resources for many methods and apparatus used in nematology even to this day.

Of Cobb's far-reaching influence on nematology research, Jenkins and Taylor write:

> *Although many workers have played important roles in development of plant nematology, none have had a greater impact, particularly in the United States, than N.A. Cobb. In 1913 Cobb published his first paper on nematology in the United States. From to 1932 he was the undisputed leader in [nematology] in this country. Through his efforts the widely renowned [USDA] nematology research program was initiated and developed. Many of his students and colleagues developed into the leaders in plant nematology in the 1930s, 1940's, and 1950's. In addition, during his productive career he contributed major discoveries in the areas of nematode taxonomy, morphology, and in methodology. Many of his techniques are still unsurpassed!*

Perhaps no one person has had as favorable an impact on the field of nematology as has Nathan Augustus Cobb.

From 1900–1925 various state-run agricultural experimental stations investigated important problems relating to agro-economy, though few stations devoted much attention to plant-parasitic nematodes. Accounts of the history of nematology (the few that exist) mention three major events occurring between 1926 and 1950 that affected the relative importance of nematodes in the eyes of farmers, legislators and the U.S. public in general. These same events had profound worldwide effects on the course of nematology research over the next fifty to seventy-five years

First, the discovery of the golden nematode in the potato fields of Long Island led to a trip by U.S. quarantine officials to the potato fields of Europe, where the devastating effects of this parasite had been known for many years. This excursion allayed all skepticism about the seriousness of this agricultural pest. Second, the introduction of the soil fumigants, D-D and EDB made available for the first time nematicides that could be used effectively and practically on a field scale. Third, the development of nematode-resistant crop cultivars brought substantial government funding to applied nematology research.

These events contributed to a shift from broad taxonomy-based nematology research to deep, yet focused investigations of plant parasitic nematodes, especially the control of agricultural pests. From the early 1930s until recently, the bulk of researchers studying nematodes have been plant pathologists by training. Consequently, nematological research leaned heavily toward answering plant pathological and agro-economical questions for the last three-quarters of the 20th century.

Notable Nematologists

- Nathan Cobb

- Michel Luc

- Maynard Jack Ramsay

- Gregor W. Yeates

Contributions to Other Sciences

Nematologists in the 1800s also contributed to other scientific fields in important ways. Butschli (1875) first observed the formation of polar bodies by nuclear subdivision in a nematode, Beneden (1883) was studying *Ascaris megalocephala* when he discovered the separation of halves of each of the chromosomes from the two parents and the mechanism of Mendelian heredity, and Boveri (1893) showed evidence for continuity of the germ plasm and that the soma may be regarded as a by-product without influence upon heredity.

Caenorhabditis elegans is a widely used model species, initially for neural development, and then for genetics. WormBase collates research on the species.

Conchology

Conchology is the study of mollusc shells. Conchology is one aspect of malacology, the study of molluscs; however, malacology is the study of molluscs as whole organisms, whereas conchology is confined to the study of their shells. It in-cludes the study of land and freshwater mollusc shells as well as seashells and extends to the study of a gastropod's operculum.

Calliostoma tigris

Shell of *Lobatus gigas*, the queen conch

Conchology is now sometimes seen as an archaic study, because relying on only one aspect of an organism's morphology can be misleading. However, a shell often gives at least some insight into molluscan taxonomy, and historically the shell was often the only part of exotic species that was available for study. Even in current museum collections it is common for the dry material (shells) to greatly exceed the amount of material that is preserved whole in alcohol.

Conchologists mainly deal with four molluscan orders: the gastropods (snails), bivalves (clams), Polyplacophora (chitons) and Scaphopoda (tusk shells). Cephalopods only have small internal shells, with the exception of the Nautiloidea. Some groups, such as the sea slug nudibranchs, have lost their shells altogether, while in others it has been replaced by a protein support structure.

Shell Collecting Versus Conchology

The terms *shell collector* and *conchologist* can be regarded as two distinct categories. Not all shell collectors are conchologists; some are primarily concerned with the aesthetic value of shells in-stead of their scientific study. It is also true that not all conchologists are shell collectors; this type of research only requires access to private or institutional shell collections. There is some debate in the conchological community, with some people regarding all shell collectors (regardless of mo-tivation) as conchologists.

A vendor in Tanzania with a variety of large seashells for sale

History

Molluscs have probably been used by primates as a food source long before humans evolved. Shell collecting, the precursor of conchology, probably goes back as far as there have been humans living near beaches.

Stone Age seashell necklaces have been found, sometimes in areas far from the ocean, indicating that they were traded. Shell jewellery is found at almost all archaeological sites, including at ancient Aztec ruins, digs in ancient China, and the Indus Valley.

During the Renaissance people began taking interest in natural objects of beauty to put in cabinets of curiosities. Because of their attractiveness, variety, durability and ubiquity, shells became a large part of these collections. Towards the end of the 17th century, people began looking at shells with scientific interest. Martin Lister in 1685–1692 published *Historia Conchyliorum*, which was the first comprehensive conchological text, having over 1000 engraved plates.

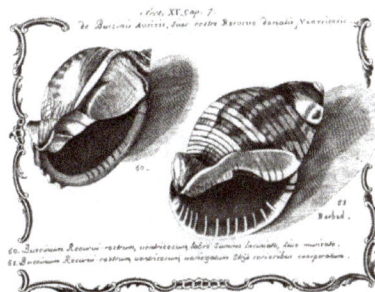

A plate from Lister's book, showing what he calls *buccinis shells*

George Rumpf, or Rumphius, (1627–1702) published the first mollusc taxonomy. He suggested "single shelled ones" (Polyplacophora, limpets, and abalone), "snails or whelks" (Gastropoda), and "two-shelled ones" (Bivalvia). Many of Rumpf's terms were adopted by Carl Linnaeus. Rumpf continued to do important scientific work after he went blind, working by touch.

The study of zoology, including conchology, was revolutionized by Swedish naturalist Carl Linnaeus and his system of binomial nomenclature. 683 of the 4000 or so animal species he described are now considered to be molluscs, although Linnaeus placed them in several phyla at the time.

There have been many prominent conchologists in the past two centuries. The Sowerby family were famous collectors, dealers, and illustrators. John Mawe (1764–1829) produced arguably the first conchology guidebook, *The Voyager's Companion or Shell-Collector's Pilot* as well as *The Linnæan System of Conchology*. Hugh Cuming (1791–1865) is famous for his huge collection and numerous discoveries of new species. Thomas Say wrote the fundamental work *American Conchology, or Descriptions of the Shells of North America, Illustrated From Coloured Figures From Original Drawings, Executed from Nature* in six volumes (1830–1834).

R. Tucker Abbott was the most prominent conchologist of the 20th century. He authored dozens of books and was museum director of the Bailey-Matthews Shell Museum, bringing the world of seashells to the public. His most prominent works are *American Seashells*, *Seashells of the World*, and *The Kingdom of the Seashell*.

Many of the finest collections of seashells are private. John DuPont and Jack Lightbourne are known for their extensive collections. John DuPont donated his shell collection to the Delaware Museum of Natural History in 1984. Emperor Hirohito of Japan also amassed a huge collection, and was a competent and respected amateur conchologist.

Museums

Many museums worldwide contain very large and scientifically important mollusc collections. However, in most cases these are research collections, behind the scenes of the museum, and thus not readily accessible to the general public in the same way that exhibits are.

The largest assemblage of mollusc shells is housed at the Smithsonian Institution which has millions of lots representing perhaps 50,000 species, versus about 35,000 species for the largest private collections.

USA

- Academy of Natural Sciences, Philadelphia

- American Museum of Natural History, New York City

- Bailey-Matthews Shell Museum in Sanibel Island, Florida: the only museum in the world dedicated entirely to shells.

- Denver Museum of Nature & Science, Denver, Colorado: approximately 17,500 shell lots.

- Museum of Comparative Zoology at Harvard, Massachusetts

- National Museum of Natural History, Washington D.C. – The Smithsonian Museum of Natural History has one of, if not the, finest shell collection in the world

Europe

- Austria, Vienna – Naturhistorisches Museum
- Belgium, Brussels – Royal Belgian Institute of Natural Sciences, one of the three largest collections
- France, Paris – Muséum national d'Histoire naturelle
- Germany, Berlin – Humboldt Museum
- Netherlands, Leiden – Natural History Museum, Leiden
- Sweden, Stockholm – Swedish Museum of Natural History
- United Kingdom
  - London – Natural History Museum
  - Manchester – the Manchester Museum has the fourth largest mollusc collection in Britain with 166,000 lots.

Organizations

Like other scientific fields, conchologists have a number of local, national, and international organizations. There are also many organizations specializing in specific subareas.

- Belgian Society for Conchology
- Club Conchylia, the German/Austrian Society for Shell Collecting
- Conchological Society of Great Britain and Ireland
- Conchologists of America
- Conquiliologistas do Brasil
- Nederlandse Malacologische Vereniging
- Unitas Malacologica

Fake Shells

Shell collectors who purchase shells from dealers may sometimes encounter shells which have been altered to represent new species or rare color varieties. It is claimed that in previous centuries, fake examples of *Epitonium scalare* were created out of rice paste.

Depictions of Shells on Stamps and Coins

Shells have been featured on over 5,000 postage stamps worldwide.

Shells have also been featured on many coins, including the Bahamian dollar (1974), the Cuban peso (1981), the Haitian gourde (1973), the Nepalese rupee (1989) and Philippine peso (1993).

Helminthology

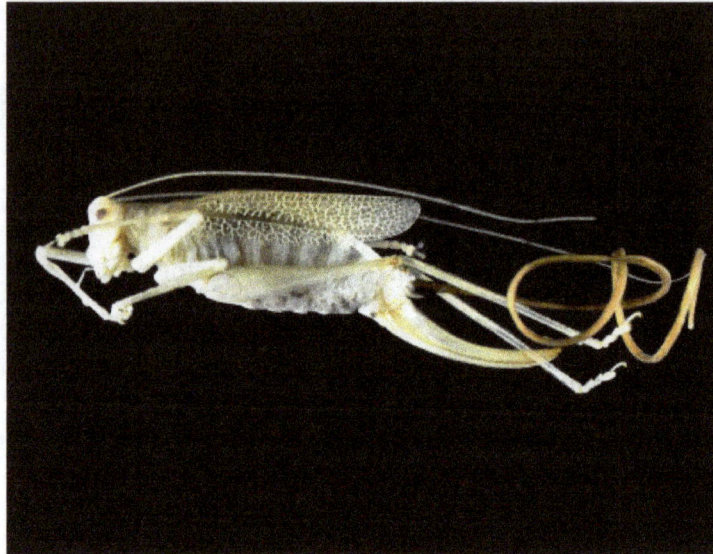

Spinochordodes parasitising *Meconema*

Helminthology is the study of parasitic worms (helminths), while helminthiasis describes the medical condition of being infected with helminths. Helminthology deals with the study of the taxonomy of helminth and the effect on their hosts.

In the 18th and early 19th century there was wave of publications on helminthology and this period has been described as the "Golden Era" of helminthology. During that period the authors Félix Dujardin, William Blaxland Benham, Peter Simon Pallas, Marcus Elieser Bloch, Otto Friedrich Müller, Johann Goeze, Friedrich Zenker, Carl Asmund Rudolphi, Otto Friedrich Bernhard von Linstow and Johann Gottfried Bremser started systematic scientific studies of the subject.

The Japanese parasitologist Satyu Yamaguti was probably one of the most active helminthologists of the 20th century; he wrote a series of six volumes entitled "Systema Helminthum".

Malacology

Malacology is the branch of invertebrate zoology that deals with the study of the Mollusca (mollusks or molluscs), the second-largest phylum of animals in terms of described species after the arthropods. Mollusks include snails and slugs, clams, octopus and squid, and numerous other kinds, many (but by no means all) of which have shells. One division of malacology, conchology, is devoted to the study of mollusk shells..

Teuthology, a branch of malacology, deals with the study of cephalopods, such as the giant squid pictured.

Fields within malacological research include taxonomy, ecology and evolution. Applied malacology studies medical, veterinary, and agricultural applications, for example mollusks as vectors of disease, as in schistosomiasis.

Archaeology employs malacology to understand the evolution of the climate, the biota of the area, and the usage of the site.

In 1681, Filippo Bonanni wrote the first book ever published that was solely about seashells, the shells of marine mollusks. The book was entitled: *Ricreatione dell' occhio e dela mente nell oservation' delle Chiociolle, proposta a' curiosi delle opere della natura, &c.* In 1868, the German Malacological Society was founded.

Zoological methods are used in malacological research. Malacological field methods and laboratory methods (such as collecting, documenting and archiving, and molecular techniques) were summarized by Sturm et al. (2006).

Malacologists

Those who study malacology are known as malacologists. Those who study primarily or exclusively the shells of mollusks are known as conchologists.

Societies

- American Malacological Society
- Association of Polish Malacologists (Stowarzyszenie Malakologów Polskich)
- Belgian Malacological Society (Société Belge de Malacologie) - French speaking

- Belgian Society for Conchology (Belgische Vereniging voor Conchyliologie) - Dutch speaking

- Conchological Society of Great Britain and Ireland

- Conchologists of America

- Dutch Malacological Society (Nederlandse Malacologische Vereniging)

- Estonian Malacological Society (Eesti Malakoloogia Ühing)

- European Quaternary Malacologists

- Freshwater Mollusk Conservation Society

- German Malacological Society (Deutsche Malakozoologische Gesellschaft)

- Hungarian Malacological Society Magyar Malakológiai Társaság

- Italian Malacological Society (Società Italiana di Malacologia)

- Malacological Society of Australasia

- Malacological Society of London

- Malacological Society of the Philippines, Inc.

- Mexican Malacological Society (Sociedad Mexicana de Malacología y Conquiliología)

- Spanish Malacological Society (Sociedad Española de Malacología)

- Western Society of Malacologists

- Brazilian Malacological Society (Sociedade Brasileira de Malacologia)

Museums

Malacological Museum in Makarska, Croatia (entrance)

Museums that have either exceptional malacological research collections (behind the scenes) and/or exceptional public exhibits of mollusks:

- Academy of Natural Sciences of Philadelphia

- American Museum of Natural History

- Bailey-Matthews Shell Museum

- Cau del Cargol Shell Museum

- Maria Mitchell Association

- Museum of Comparative Zoology at Harvard

- Rinay

- Royal Belgian Institute of Natural Sciences, Brussels: with a collection of more than 9 million shells (mainly from the collection of Philippe Dautzenberg)

- Smithsonian Institution

Herpetology

Herpetology (from Greek "herpien" meaning "to creep") is the branch of zoology concerned with the study of amphibians (including frogs, toads, salamanders, newts, and caecilians (gymnophiona)) and reptiles (including snakes, lizards, amphisbaenids, turtles, terrapins, tortoises, crocodilians, and the tuataras). Batrachology is a further subdiscipline of herpetology concerned with the study of amphibians alone.

Male golden toad

Herpetology is concerned with poikilothermic, ectothermic tetrapods. Under this definition "herps" (or sometimes "herptiles" or "herpetofauna") exclude fish, but it is not uncommon for herpetological and ichthyological scientific societies to "team up", publishing joint journals and holding conferences in order to foster the exchange of ideas between the fields. One of the most prestigious organizations, the American Society of Ichthyologists and Herpetologists, is an example of this. Many herpetological societies exist today, having been formed to promote interest in reptiles and amphibians both captive and wild.

Herpetology offers benefits to humanity in the study of the role of amphibians and reptiles in global ecology, especially because amphibians are often very sensitive to environmental changes, offering a visible warning to humans that significant changes are taking place. Some toxins and venoms produced by reptiles and amphibians are useful in human medicine. Currently, some snake venom has been used to create anti-coagulants that work to treat stroke victims and heart-attack cases.

Etymology

"Herp" is a vernacular term for reptiles and amphibians. It is derived from the old term "herpetile", with roots back to Linnaeus's classification of animals, in which he grouped reptiles and amphibians together in the same class. There are over 6700 species of amphibians and over 9000 species of reptiles. In spite of its modern taxonomic irrelevance, the term has persisted, particularly in the names of herpetology, the scientific study of reptiles and amphibians, and herpetoculture, the captive care and breeding of reptiles and amphibians.

Careers

Career options in the field of herpetology include, but are not limited to lab research, field studies and survey, zoological staff, museum staff and college teaching.

In modern academic science, it is rare for individuals to consider themselves a herpetologist first and foremost. Most individuals focus on a particular field such as ecology, evolution, taxonomy, physiology, or molecular biology, and within that field ask questions pertaining to or best answered by examining reptiles and amphibians. For example, an evolutionary biologist who is also a herpetologist may choose to work on an issue such as evolution of warning coloration in coral snakes.

Modern herpetological writers of note include Mark O'Shea and Philip Purser. Modern herpetological showmen of note include Jeff Corwin, Steve Irwin, popularly known as the "Crocodile Hunter", and the star Austin Stevens, popularly known as 'AustinSnakeman' in the TV series Austin Stevens: Snakemaster.

Study

Most colleges or universities do not offer a major in herpetology at the undergraduate or even the graduate level. Instead, persons interested in herpetology select a major in the biological sciences. The knowledge learned about all aspects of the biology of animals is then applied to an individual study of herpetology.

References

- Prosser, C. Ladd (1991). Comparative Animal Physiology, Environmental and Metabolic Animal Physiology (4th ed.). Hoboken, NJ: Wiley-Liss. pp. 1–12. ISBN 0-471-85767-X.

- Hall, John (2011). Guyton and Hall textbook of medical physiology (12th ed.). Philadelphia, Pa.: Saunders/ Elsevier. p. 3. ISBN 978-1-4160-4574-8.

- Feder, ME; Bennett, AF; WW, Burggren; Huey, RB (1987). New directions in ecological physiology. New York:

Cambridge University Press. ISBN 978-0-521-34938-3.

- McGreevy, Paul; Robert Boakes (2011). Carrots and Sticks: Principles of Animal Training. Darlington Press. pp. xi–23. ISBN 978-1-921364-15-0. Retrieved 9 September 2016.

- Bourg, Julian (2007). From Revolution to Ethics: May 1968 and Contemporary French Thought. McGill-Queen's Press - MQUP. p. 155. ISBN 978-0-7735-7621-6.

- Bateson, P. (1991). The Development and Integration of Behaviour: Essays in Honour of Robert Hinde. Cambridge University Press. p. 479. ISBN 978-0-521-40709-0.

- Keil, Frank C.; Robert Andrew Wilson (2001). The MIT encyclopedia of the cognitive sciences. MIT Press. p. 184. ISBN 978-0-262-73144-7.

- Buchmann, Stephen (2006). Letters from the Hive: An Intimate History of Bees, Honey, and Humankind. Random House of Canada. p. 105. ISBN 978-0-553-38266-2.

- Mercer, Jean (2006). Understanding attachment: parenting, child care, and emotional development. Greenwood Publishing Group. p. 19. ISBN 978-0-275-98217-1.

- Hoppitt, W.; Laland, K.N. (2013). Social Learning: An Introduction to Mechanisms, Methods, and Models. Princeton University Press. ISBN 978-1-4008-4650-4.

- Cummings, Mark; Carolyn Zahn-Waxler; Ronald Iannotti (1991). Altruism and aggression: biological and social origins. Cambridge University Press. p. 7. ISBN 978-0-521-42367-0.

- Liddell, Henry George and Robert Scott (1980). A Greek-English Lexicon (Abridged Edition). United Kingdom: Oxford University Press. ISBN 0-19-910207-4.

- Chapman, A. D. (2006). Numbers of living species in Australia and the World. Canberra: Australian Biological Resources Study. pp. 60pp. ISBN 978-0-642-56850-2.

- Antonio Saltini, Storia delle scienze agrarie, 4 vols, Bologna 1984-89, ISBN 88-206-2412-5, ISBN 88-206-2413-3, ISBN 88-206-2414-1, ISBN 88-206-2415-X

- Haviland, William A.; Prins, Harald E. L.; McBride, Bunny; Walrath, Dana (2010), Cultural Anthropology: The Human Challenge (13th ed.), Cengage Learning, ISBN 0-495-81082-7

- Kottak, Conrad Phillip (2010). Anthropology : appreciating human diversity (14th ed.). New York: McGraw-Hill. pp. 579–584. ISBN 978-0-07-811699-5.

- Townsend, Patricia K. (2009). Environmental anthropology : from pigs to policies (2nd ed.). Prospect Heights, Ill.: Waveland Press. p. 104. ISBN 978-1-57766-581-6.

- Melissa Checker (August 2005). Polluted promises: environmental racism and the search for justice in a southern town. NYU Press. ISBN 978-0-8147-1657-1. Retrieved 3 April 2011.

- Goodman, Alan H.; Thomas L. Leatherman (eds.) (1998). Building A New Biocultural Synthesis. University of Michigan Press. ISBN 978-0-472-06606-3.

- Adrian Franklin (20 September 1999). Animals and Modern Cultures: A Sociology of Human-Animal Relations in Modernity. SAGE Publications. ISBN 978-0-7619-5623-5.

- Schiebinger, Londa (2001). Has Feminism Changed Science?. First Harvard University Press. p. 129. ISBN 0-674-00544-9.

- Sutherland, W. J., Newton, Ian and Green, Rhys (2004). Bird ecology and conservation: a handbook of techniques. Oxford University Press. ISBN 0-19-852086-7.

- Stresemann, Erwin (1975). Ornithology. From Aristotle to the Present. Cambridge, Massachusetts: Harvard University Press. pp. 170–191. ISBN 0-674-64485-9.

Understanding Vertebrates and Invertebrates

Animals that possess backbones and chordates are known as vertebrates and those without them are invertebrates. Vertebrates evolved from invertebrate species in the Cambrian and Devonian period of the Earth's history. Binomial classification, or naming of species relies on such anatomical differentiation among animals. The basic classification of zoology is dealt with in this chapter.

Vertebra

In the vertebrate spinal column, each vertebra is an irregular bone with a complex structure composed of bone and some hyaline cartilage, the proportions of which vary according to the segment of the backbone and the species of vertebrate.

The basic configuration of a vertebra varies; the large part is the body, and the central part is the centrum. The upper and lower surfaces of the vertebra body give attachment to the intervertebral discs. The posterior part of a vertebra forms a vertebral arch, in eleven parts, consisting of two pedicles, two laminae, and seven processes. The laminae give attachment to the ligamenta flava (ligaments of the spine). There are vertebral notches formed from the shape of the pedicles, which form the intervertebral foramina when the vertebrae articulate. These foramina are the entry and exit conducts for the spinal nerves. The body of the vertebra and the vertebral arch form the vertebral foramen, the larger, central opening that accommodates the spinal canal, which encloses and protects the spinal cord.

Vertebrae articulate with each other to give strength and flexibility to the spinal column, and the shape at their back and front aspects determines the range of movement. Structurally, vertebrae are essentially alike across the vertebrate species, with the greatest difference seen between an aquatic animal and other vertebrate animals. As such, vertebrates take their name from the vertebrae that compose the vertebral column.

Structure

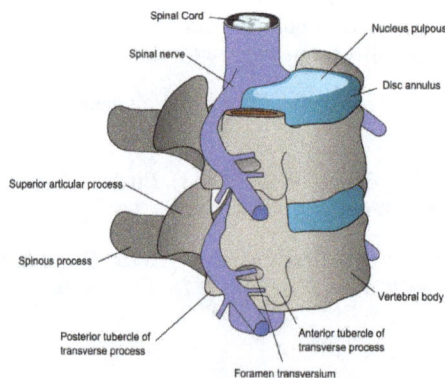

Side view of vertebrae

Each vertebra is an irregular bone. The size of the vertebrae varies according to placement in the vertebral column, spinal loading, posture and pathology. Along the length of the spine the vertebrae change to accommodate different needs related to stress and mobility.

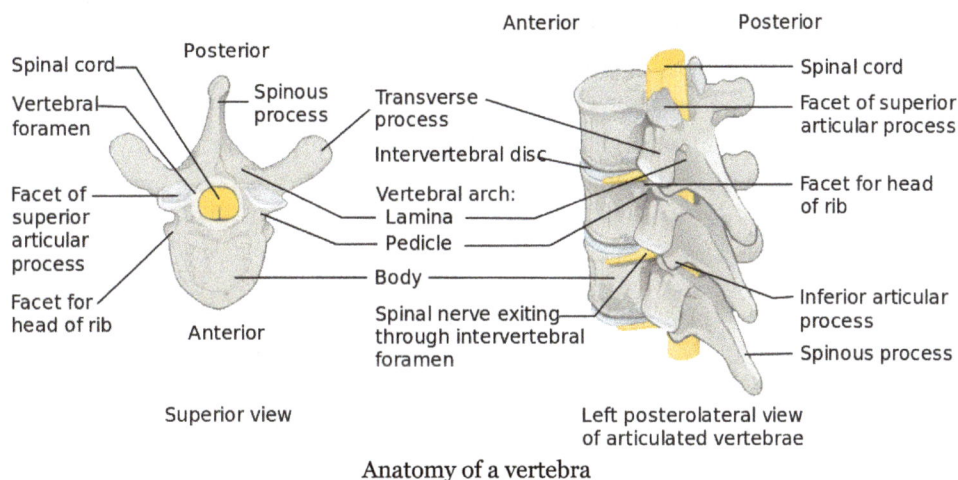

Anatomy of a vertebra

Every vertebra has a body, which consists of a large anterior middle portion called the centrum and a posterior vertebral arch, also called a neural arch. The body is composed of cancellous bone, which is the spongy type of osseous tissue, whose micro-anatomy has been specifically studied within the pedicle bones. This cancellous bone is in turn, covered by a thin coating of cortical bone (or compact bone), the hard and dense type of osseous tissue. The vertebral arch and processes have thicker coverings of cortical bone. The upper and lower surfaces of the body of the vertebra are flattened and rough in order to give attachment to the intervertebral discs. These surfaces are the vertebral endplates which are in direct contact with the intervertebral discs and form the joint. The endplates are formed from a thickened layer of the cancellous bone of the vertebral body, the top layer being more dense. The endplates function to contain the adjacent discs, to evenly spread the applied loads, and to provide anchorage for the collagen fibers of the disc. They also act as a semi-permeable interface for the exchange of water and solutes.

The vertebral arch is formed by pedicles and laminae. Two pedicles extend from the sides of the vertebral body to join the body to the arch. The pedicles are short thick processes that extend, one from each side, posteriorly, from the junctions of the posteriolateral surfaces of the centrum, on its upper surface. From each pedicle a broad plate, a lamina, projects backwards and medialwards to join and complete the vertebral arch and form the posterior border of the vertebral foramen, which completes the triangle of the vertebral foramen. The upper surfaces of the laminae are rough to give attachment to the ligamenta flava. These ligaments connect the laminae of adjacent vertebra along the length of the spine from the level of the second cervical vertebra. Above and below the pedicles are shallow depressions called vertebral notches (*superior* and *inferior*). When the vertebrae articulate the notches align with those on adjacent vertebrae and these form the openings of the intervertebral foramina. The foramina allow the entry and exit of the spinal nerves from each vertebra, together with associated blood vessels. The articulating vertebrae provide a strong pillar of support for the body.

There are seven processes projecting from the vertebra; a spinous process, two transverse processes, and four articular processes. A major part of a vertebra is a backward extending spinous process which projects centrally. In humans this process points backward and downward from the junction of the laminae, but in those animals without an erect stance they are directed upwards. The spinous process serves to attach muscles and ligaments.

The two transverse processes, one on each side of the vertebral body project from either side at the point where the lamina joins the pedicle, between the superior and inferior articular processes. They also serve for the attachment of muscles and ligaments, in particular the intertransverse ligaments. There is a facet on each of the transverse processes of thoracic vertebrae which articulates with the tubercle of the rib. A facet on each side of the thoracic vertebral body articulates with the head of the rib. There are superior and inferior articular facet joints on each side of the vertebra, which serve to restrict the range of movement possible. These facets are joined by a thin portion of the vertebral arch called the *pars interarticularis*.

The transverse process of a lumbar vertebra is also sometimes called the **costal** or **costiform process** because it corresponds to a rudimentary rib (*costa*) which, as opposed to the thorax, is not developed in the lumbar region.

Variation in the Human

Vertebrae take their names from the regions of the vertebral column that they occupy. There are thirty-three vertebrae in the human vertebral column—seven cervical vertebrae, twelve thoracic vertebrae, five lumbar vertebrae, five fused sacral vertebrae forming the sacrum and three to five coccygeal vertebrae, forming the coccyx. The regional vertebrae increase in size as they progress downwards but become smaller in the coccyx.

Cervical Vertebrae

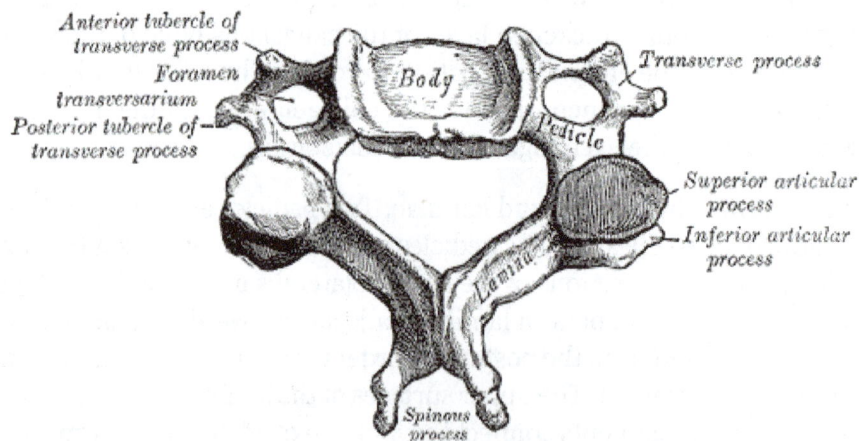

A typical cervical vertebra

There are seven cervical vertebrae (but eight cervical spinal nerves), designated C1 through C7. These bones are, in general, small and delicate. Their spinous processes are short (with the exception of C2 and C7, which have palpable spinous processes). C1 is also called the atlas, and C2 is also called the axis. The structure of these vertebrae is the reason why the neck and head have a large range of motion. The atlanto-occipital joint allows the skull to move up and down, while the atlanto-axial joint allows the upper neck to twist left and right. The axis also sits upon the first intervertebral disc of the spinal column.

Cervical vertebrae possess transverse foramina to allow for the vertebral arteries to pass through on their way to the foramen magnum to end in the circle of Willis. These are the smallest, lightest vertebrae and the vertebral foramina are triangular in shape. The spinous processes are short and often bifurcated (the spinous process of C7, however, is not bifurcated, and is substantially longer than that of the other cervical spinous processes).

The atlas differs from the other vertebrae in that it has no body and no spinous process. It has instead a ring-like form, having an anterior and a posterior arch and two lateral masses. At the outside centre points of both arches there is a tubercle; an anterior tubercle and a posterior tubercle for the attachment of muscles. The front surface of the anterior arch is convex and its anterior tubercle gives attachment to the longus colli muscle. The posterior tubercle is a rudimentary spinous process and gives attachment to the rectus capitis posterior minor muscle. The spinous process is small so as not to interfere with the movement between the atlas and the skull. On the under surface is a facet for articulation with the dens of the axis.

Specific to the cervical vertebra is the transverse foramen (also known as *foramen transversarium*). This is an opening on each of the transverse processes which gives passage to the vertebral artery and vein and a sympathetic nerve plexus. On the cervical vertebrae other than the atlas, the anterior and posterior tubercles are on either side of the transverse foramen on each transverse process. The anterior tubercle on the sixth cervical vertebra is called the carotid tubercle because it separates the carotid artery from the vertebral artery.

There is a hook-shaped uncinate process on the side edges of the top surface of the bodies of the third to the seventh cervical vertebrae, and also of the first thoracic vertebra. Together with the vertebral disc, this uncinate process prevents a vertebra from sliding backwards off the vertebra below it and limits lateral flexion (side-bending). Luschka's joints involve the vertebral uncinate processes.

The spinous process on C7 is distinctively long and gives the name vertebra prominens to this vertebra. Also a cervical rib can develop from C7 as an anatomical variation.

The term cervicothoracic is often used to refer to the cervical and thoracic vertebrae together, and sometimes also their surrounding areas.

Thoracic Vertebrae

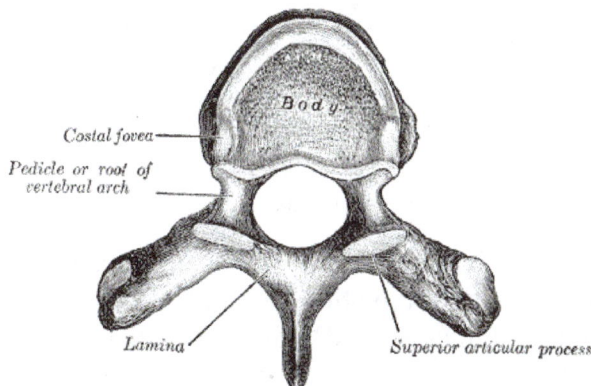

A typical thoracic vertebra

The twelve thoracic vertebrae and their transverse processes have surfaces that articulate with the ribs. Some rotation can occur between the thoracic vertebrae, but their connection with the rib cage prevents much *flexion* or other movement. They may also be known as 'dorsal vertebrae', in the human context.

The vertebral bodies are roughly heart-shaped and are about as wide anterio-posterioly as they are in the transverse dimension. Vertebral foramina are roughly circular in shape.

The top surface of the first thoracic vertebra has a hook-shaped uncinate process, just like the cervical vertebrae.

The term thoracolumbar is sometimes used to refer to the thoracic and lumbar vertebrae together, and sometimes also their surrounding areas.

The thoracic vertebrae attach to ribs and so have articular facets specific to them; these are the superior, transverse and inferior costal facets. As the vertebrae progress down the spine they increase in size to match up with the adjoining lumbar section.

Lumbar Vertebrae

Lumbar vertebra.

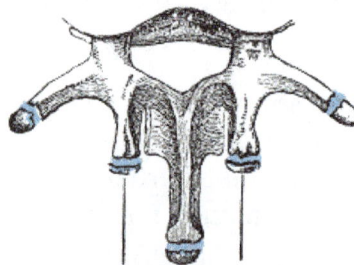

2 additional centers for mammillary processes

Lumbar vertebra showing mammillary processes

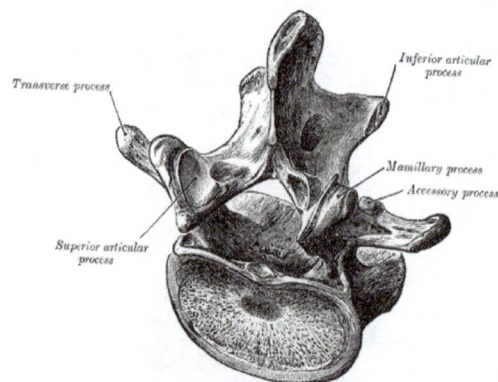

A typical lumbar vertebra

The five lumbar vertebrae are the largest of the vertebrae and are very robust in construction, as they need to support more weight than the other vertebrae. They allow significant *flexion, extension* and moderate lateral flexion (side-bending). The discs between these vertebrae create a natural lumbar lordosis (a spinal curvature that is concave posteriorly).This is due to the difference in thickness between the front and back parts of the intervertebral discs.

The lumbar vertebrae are located between the ribcage and the pelvis and are the largest of the vertebrae. The pedicles are strong as are the laminae and the spinous process is thick and broad. The vertebral foramen is large and triangular. The transverse processes are long and narrow and three tubercles can be seen on them. These are a lateral cosiform process, a mammillary process and an accessory process. The superior, or upper tubercle is the mammillary process which connects with the superior articular process. The multifidus muscle attaches to the mammillary process and this muscle extends through the length of the vertebral column, giving support. The inferior, or lower tubercle is the accessory process and this is found at the back part of the base of the transverse process. The term lumbosacral is often used to refer to the lumbar and sacral vertebrae together, and sometimes includes their surrounding areas.

Sacrum

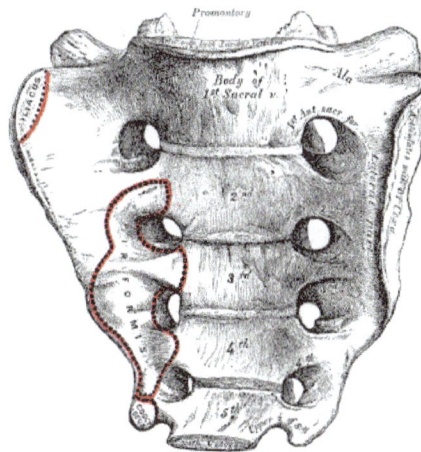

Sacrum

There are five sacral vertebrae (S1-S5) which are fused in maturity, into one large bone, the sacrum, with no intervertebral discs. The sacrum with the ilium forms a sacroiliac joint on each side of the pelvis, which articulates with the hips.

Coccyx

The last three to five coccygeal vertebrae (but usually four) (Co1-Co5) make up the tailbone or coccyx. There are no intervertebral discs.

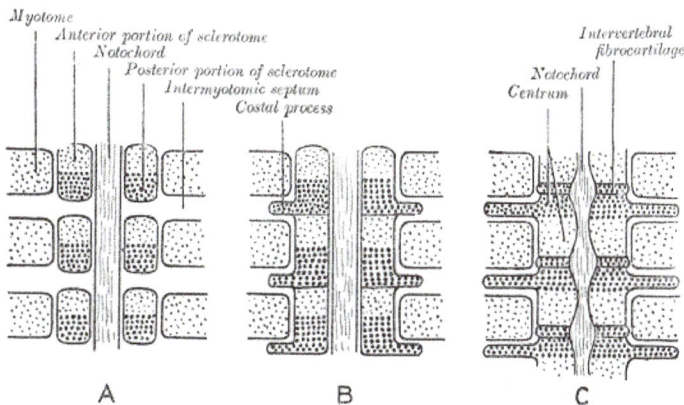

A   B   C

Somites form in the early embryo and some of these develop into sclerotomes. The sclerotomes form the vertebrae as well as the rib cartilage and part of the occipital bone. From their initial location within the somite, the sclerotome cells migrate medially towards the notochord. These cells meet the sclerotome cells from the other side of the paraxial mesoderm. The lower half of one sclerotome fuses with the upper half of the adjacent one to form each vertebral body. From this vertebral body, sclerotome cells move dorsally and surround the developing spinal cord, forming the vertebral arch. Other cells move distally to the costal processes of thoracic vertebrae to form the ribs.

Function

Functions of vertebrae include:

1. Support. The vertebrae function in the skeletomuscular system by forming the vertebral column to support the body.

2. Protection. Vertebrae contain a vertebral foramen for the passage of the spinal canal and its enclosed spinal cord and covering meninges. They also afford sturdy protection for the spinal cord. The upper and lower surfaces of the centrum are flattened and rough in order to give attachment to the intervertebral discs.

3. Movement. The vertebrae also provide the openings, the intervertebral foramina which allow the entry and exit of the spinal nerves. Similarly to the surfaces of the centrum, the upper and lower surfaces of the fronts of the laminae are flattened and rough to give attachment to the ligamenta flava. Working together in the vertebral column their sections provide controlled movement and flexibility.

4. Feeding the Intervertebral discs, through the reflex (hyaline ligament) plate that separates the cancellous bone of the vertebral body from each disk.

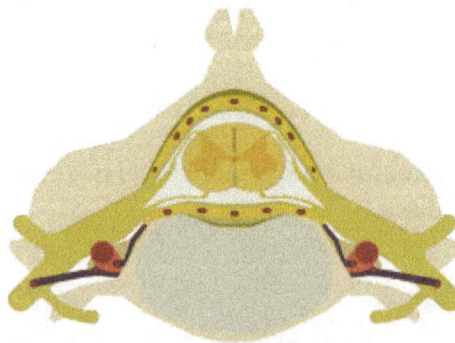

The spinal cord nested in the vertebral column.

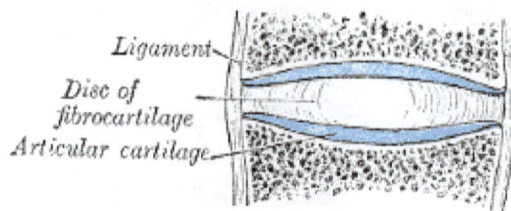

Vertebral joint

Other animals

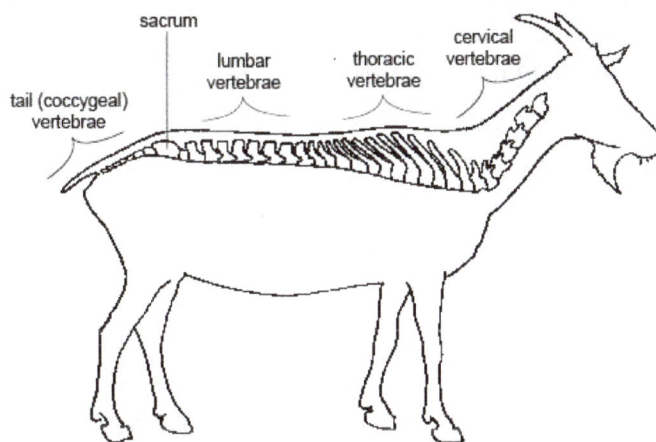

Regions of vertebrae in the goat

In other animals the vertebrae take the same regional names except for the coccygeal – in animals with tails the separate vertebrae are usually called the caudal vertebrae. Because of the different types of locomotion and support needed between the aquatic and other vertebrates, the vertebrae between them show the most variation, though basic features are shared. The spinous processes which are backward extending are directed upwards in animals without an erect stance. These processes can be very large in the larger animals as they attach to the muscles and ligaments of the body. In the elephant the vertebrae are connected by tight joints, which limit the backbone's flexibility. Spinous processes are exaggerated in some animals, such as the extinct *Dimetrodon* and *Spinosaurus*, where they form a sailback or finback.

Vertebrae with saddle-shaped articular surfaces on their bodies, are called "heterocoelous", which allows vertebrae to flex both vertically and horizontally, while preventing twisting motions. Such vertebrae are found in the necks of birds and some turtles.

In many species, though not in mammals, the cervical vertebrae bear ribs. In many groups, such as lizards and saurischian dinosaurs, the cervical ribs are large; in birds, they are small and completely fused to the vertebrae. The transverse processes of mammals are homologous to the cervical ribs of other amniotes. In the whale the cervical vertebrae are typically fused, an adaptation trading flexibility for stability during swimming. All mammals except manatees and sloths have seven cervical vertebrae, whatever the length of the neck. This includes seemingly unlikely animals such as the giraffe, the camel, and the blue whale, for example. Birds usually have more cervical vertebrae with most having a highly flexible neck consisting of 13-25 vertebrae.

In all mammals the thoracic vertebrae are connected to ribs and their bodies differ from the other regional vertebrae due to the presence of facets. Each vertebra has a facet on each side of the vertebral body which articulates with the head of a rib. There is also a facet on each of the transverse processes which articulates with the tubercle of a rib. The number of thoracic vertebrae varies considerably across the species. Most marsupials have thirteen, but koalas only have eleven. The norm is twelve to fifteen in mammals, (twelve in the human), though there are from eighteen to twenty in the horse, tapir, rhinoceros and elephant. In certain sloths there is an extreme number of twenty-five and at the other end only nine in the cetacean.

There are fewer lumbar vertebrae in the *Pan* genus species of chimpanzees and gorillas, which have three in contrast to the five in the *Homo*. This reduction in number gives an inability of the lumbar spine to lordose but gives an anatomy that favours vertical climbing, and hanging ability more suited to feeding locations in high-canopied regions. The bonobo differs by having four lumbar vertebrae.

Caudal vertebrae are the bones that make up the tails of vertebrates. They range in number from a few to fifty, depending on the length of the animal's tail. In humans and other tailless primates, they are called the coccygeal vertebrae, number from three to five and are fused into the coccyx.

Clinical Significance

There are a number of congenital vertebral anomalies, mostly involving variations in the shape or number of vertebrae, and many of which are unproblematic. Others though can cause compression of the spinal cord. Wedge-shaped vertebrae, called *hemivertebrae* can cause an angle to form in the spine which can result in the spinal curvature diseases of kyphosis, scoliosis and lordosis. Severe cases can cause spinal cord compression. Block vertebrae where some vertebrae have become fused can cause problems. Spina bifida can result from the incomplete formation of the vertebral arch.

Spondylolysis is a defect in the pars interarticularis of the vertebral arch. In most cases this occurs in the lowest of the lumbar vertebrae (L5), but may also occur in the other lumbar vertebrae, as well as in the thoracic vertebrae.

Spinal disc herniation, more commonly called a *slipped disc*, is the result of a tear in the outer ring (anulus fibrosus) of the intervertebral disc, which lets some of the soft gel-like material, the nucleus pulposus, bulge out in a hernia. This may be treated by a minimally-invasive endoscopic procedure called Tessys method.

A laminectomy is a surgical operation to remove the laminae in order to access the spinal canal. The removal of just part of a lamina is called a laminotomy.

A pinched nerve caused by pressure from a disc, vertebra or scar tissue might be remedied by a foraminotomy to broaden the intervertebral foramina and relieve pressure. It can also be caused by a foramina stenosis, a narrowing of the nerve opening, as a result of arthritis.

Another condition is spondylolisthesis when one vertebra slips forward onto another. The reverse of this condition is retrolisthesis where one vertebra slips backwards onto another.

The vertebral pedicle is often used as a radiographic marker and entry point in vertebroplasty, kyphoplasty, and spinal fusion procedures.

The arcuate foramen is a common anatomical variation more frequently seen in females. It is a bony *bridge* found on the first cervical vertebra, the atlas where it covers the groove for the vertebral artery.

Degenerative disc disease is a condition usually associated with ageing in which one or more discs degenerate. This can often be a painfree condition but can also be very painful.

Vertebrate

Vertebrates comprise all species of animals within the subphylum Vertebrata (chordates with backbones). Vertebrates represent the overwhelming majority of the phylum Chordata, with currently about 64,000 species described. Vertebrates include the jawless fish and the jawed vertebrates, which include the cartilaginous fish (sharks and rays) and the bony fish.

A bony fish clade known as the lobe-finned fishes is included with tetrapods, which are further divided into amphibians, reptiles, mammals, and birds. Extant vertebrates range in size from the frog species *Paedophryne amauensis*, at as little as 7.7 mm (0.30 in), to the blue whale, at up to 33 m (108 ft). Vertebrates make up less than five percent of all described animal species; the rest are invertebrates, which lack vertebral columns.

The vertebrates traditionally include the hagfish, which do not have proper vertebrae, though their closest living relatives, the lampreys, do. Hagfish do, however, possess a cranium. For this reason, the vertebrate subphylum is sometimes referred to as "Craniata" when discussing morphology.

Molecular analysis since 1992 has suggested that the hagfish are most closely related to lampreys, and so also are vertebrates in a monophyletic sense. Others consider them a sister group of vertebrates in the common taxon of craniata.

Etymology

The word origin of *vertebrate* derives from the Latin word *vertebratus* (Pliny), meaning *joint of the spine*. The Proto-Indo-European language origins are still unclear.

Vertebrate is closely related to the word *vertebra*, which refers to any of the bones or segments of the spinal column.

Anatomy and Morphology

All vertebrates are built along the basic chordate body plan: a stiff rod running through the length of the animal (vertebral column or notochord), with a hollow tube of nervous tissue (the spinal cord) above it and the gastrointestinal tract below.

In all vertebrates, the mouth is found at, or right below, the anterior end of the animal, while the anus opens to the exterior before the end of the body. The remaining part of the body continuing after of the anus forms a tail with vertebrae and spinal cord, but no gut.

Vertebral Column

The defining characteristic of a vertebrate is the vertebral column, in which the notochord (a stiff rod of uniform composition) found in all chordates has been replaced by a segmented series of stiffer elements (vertebrae) separated by mobile joints (intervertebral discs, derived embryonically and evolutionarily from the notochord).

However, a few vertebrates have secondarily lost this anatomy, retaining the notochord into adulthood, such as the sturgeon and *Latimeria*. Jawed vertebrates are typified by paired appendages (fins or legs, which may be secondarily lost), but this is not part of the definition of vertebrates as a whole.

Fossilized skeleton of *Diplodocus carnegii*, showing an extreme example of the backbone that characterizes the vertebrates.

Gills

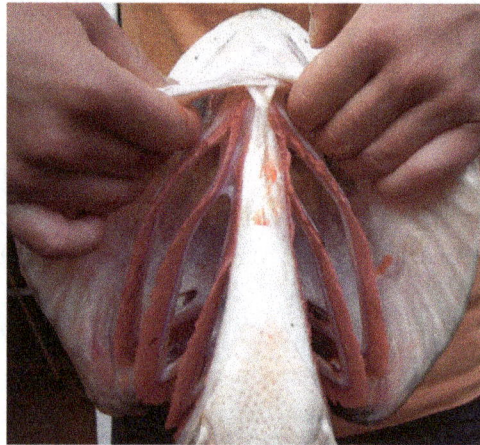

Gill arches bearing gills in a pike

All basal vertebrates breathe with gills. The gills are carried right behind the head, bordering the posterior margins of a series of openings from the pharynx to the exterior. Each gill is supported by a cartilagenous or bony gill arch. The bony fish have three pairs of arches, cartilaginous fish have five to seven pairs, while the primitive jawless fish have seven. The vertebrate ancestor no doubt had more arches, as some of their chordate relatives have more than 50 pairs of gills.

In amphibians and some primitive bony fishes, the larvae bear external gills, branching off from the gill arches. These are reduced in adulthood, their function taken over by the gills proper in fishes and by lungs in most amphibians. Some amphibians retain the external larval gills in adulthood, the complex internal gill system as seen in fish apparently being irrevocably lost very early in the evolution of tetrapods.

While the higher vertebrates do not have gills, the gill arches form during fetal development, and lay the basis of essential structures such as jaws, the thyroid gland, the larynx, the *columella* (corresponding to the stapes in mammals) and in mammals the malleus and incus.

Central Nervous System

The central nervous system of vertebrates is based on a hollow nerve cord running along the length of the animal. Of particular importance and unique to vertebrates is the presence of neural crest cells. These are progenitors of stem cells, and critical to coordinating the functions of cellular components. Neural crest cells migrate through the body from the nerve cord during development, and initiate the formation of neural ganglia and structures such as the jaws and skull.

The vertebrates are the only chordate group to exhibit cephalisation, the concentration of brain functions in the head. A slight swelling of the anterior end of the nerve cord is found in the lancelet, though it lacks the eyes and other complex sense organs comparable to those of vertebrates. Other chordates do not show any trends towards cephalisation.

A peripheral nervous system branches out from the nerve cord to innervate the various systems. The front end of the nerve tube is expanded by a thickening of the walls and expansion of the central canal of spinal cord into three primary brain vesicles: The prosencephalon (forebrain), mesencephalon (midbrain) and rhombencephalon (hindbrain), further differentiated in the various vertebrate groups. Two laterally placed eyes form around outgrowths from the midbrain, except in hagfish, though this may be a secondary loss. The forebrain is well developed and subdivided in most tetrapods, while the midbrain dominate in many fish and some salamanders. Vesicles of the forebrain are usually paired, giving rise to hemispheres like the cerebral hemispheres in mammals.

The resulting anatomy of the central nervous system, with a single hollow nerve cord topped by a series of (often paired) vesicles, is unique to vertebrates. All invertebrates with well-developed brains, such as insects, spiders and squids, have a ventral rather than dorsal system of ganglions, with a split brain stem running on each side of the mouth or gut.

Evolutionary History

First Vertebrates

The early vertebrate *Haikouichthys*

Vertebrates originated about 525 million years ago during the Cambrian explosion, which saw the rise in organism diversity. The earliest known vertebrate is believed to be the *Myllokunmingia*. Another early vertebrate is *Haikouichthys ercaicunensis*. Unlike the other fauna that dominated the Cambrian, these groups had the basic vertebrate body plan: a notochord, rudimentary vertebrae, and a well-defined head and tail. All of these early vertebrates lacked jaws in the common sense and relied on filter feeding close to the seabed. A vertebrate group of uncertain phylogeny, small-eel-like conodonts, are known from microfossils of their paired tooth segments from the late Cambrian to the end of the Triassic.

From fish to amphibians

Acanthostega, a fish-like early labyrinthodont.

The first jawed vertebrates appeared in the latest Ordovician and became common in the Devonian, often known as the "Age of Fishes". The two groups of bony fishes, the actinopterygii and sarcopterygii, evolved and became common. The Devonian also saw the demise of virtually all jawless fishes, save for lampreys and hagfish, as well as the Placodermi, a group of armoured fish that dominated the entirety of that period since the late Silurian. The Devonian also saw the rise of the first labyrinthodonts, which was a transitional form between fishes and amphibians.

Mesozoic Vertebrates

The Amniotes branched from labyrinthodonts in the subsequent Carboniferous period. The Parareptilia and synapsid amniotes were common during the late Paleozoic, while the diapsids became dominant during the Mesozoic. In the sea, the bony fishes became dominant. The birds a derived form of dinosaurs evolved in the Jurassic. The demise of the non avian dinosaurs at the end of the Cretaceous allowed for expansion of the mammals, which had evolved from the therapsids, a group of synapsid amniotes, during the late Triassic Period.

After the Mesozoic

The Cenozoic world has seen great diversification of bony fishes, frogs, birds and mammals.

Over half of all living vertebrate species (about 32,000 species) are fishes (non-tetrapod craniates), a diverse set of lineages that inhabit all the world's aquatic ecosystems, from snow minnows (Cypriniformes) in Himalayan lakes at elevations over 4,600 metres (15,100 feet) to flatfishes (order Pleuronectiformes) in the Challenger Deep, the deepest ocean trench at about 11,000 metres (36,000 feet). Fishes of myriad varieties are the main predators in most of the world's water bodies, both freshwater and marine. The rest of the vertebrate species are tetrapods, a single lineage that includes amphibians (with roughly 7,000 species); mammals (with approximately 5,500 species); and reptiles and birds (with about 20,000 species divided evenly between the two classes). Tetrapods comprise the dominant megafauna of most terrestrial environments and also include many partially or fully aquatic groups (e.g., sea snakes, penguins, cetaceans).

Classification

There are several ways of classifying animals. Evolutionary systematics relies on anatomy, physiology and evolutionary history, which is determined through similarities in anatomy and, if possible, the genetics of organisms. Phylogenetic classification is based solely on phylogeny. Evolutionary systematics gives an overview; phylogenetic systematics gives detail. The two systems are thus complementary rather than opposed.

Traditional Classification

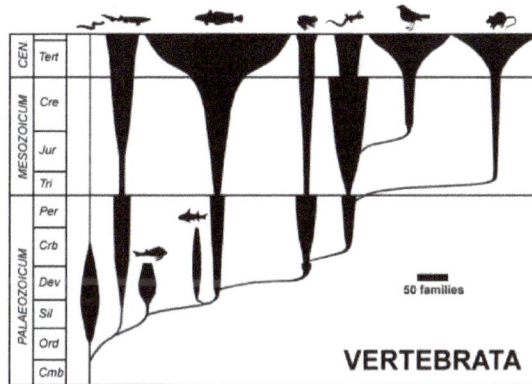

Traditional spindle diagram of the evolution of the vertebrates at class level

Conventional classification has living vertebrates grouped into seven classes based on traditional interpretations of gross anatomical and physiological traits. This classification is the one most commonly encountered in school textbooks, overviews, non-specialist, and popular works. The extant vertebrates are:

- Subphylum Vertebrata

    o Class Agnatha (jawless fishes)

    o Class Chondrichthyes (cartilaginous fishes)

    o Class Osteichthyes (bony fishes)

    o Class Amphibia (amphibians)

    o Class Reptilia (reptiles)

    o Class Aves (birds)

    o Class Mammalia (mammals)

In addition to these comes two classes of extinct armoured fishes, the Placodermi and the Acanthodii. Other ways of classifying the vertebrates have been devised, particularly with emphasis on the phylogeny on early amphibians and reptiles. An example based on Janvier (1981, 1997), Shu *et al.* (2003), and Benton (2004) is given here:

- Subphylum Vertebrata

    o *Palaeospondylus*

    o Superclass Agnatha or Cephalaspidomorphi (lampreys and other jawless fishes)

    o Infraphylum Gnathostomata (vertebrates with jaws)

- Class †Placodermi (extinct armoured fishes)

- Class Chondrichthyes (cartilaginous fishes)

- Class †Acanthodii (extinct spiny "sharks")

- Superclass Osteichthyes (bony vertebrates)

    - Class Actinopterygii (ray-finned bony fishes)

    - Class Sarcopterygii (lobe-finned fishes, tetrapods are inside this class)

    - Class Amphibia (amphibians, some ancestral to the amniotes)- now a paraphyletic group

    - Class Synapsida (mammals are placed inside this thought to be extinct taxon)

    - Class Sauropsida (reptiles, birds are inside this group in a monophyletic way)

While this traditional classification is orderly, most of the groups are paraphyletic, i.e. do not contain all descendants of the class' common ancestor. For instance, descendants of the first reptiles include modern reptiles as well as mammals and birds. Most of the classes listed are not "complete" (and therefore paraphyletic) taxa, meaning they do not include all the descendants of the first representative of the group. For example, the agnathans have given rise to the jawed vertebrates; the bony fishes have given rise to the land vertebrates; the traditional "amphibians" have given rise to the reptiles (traditionally including the synapsids, or mammal-like "reptiles"), which in turn have given rise to the mammals and birds. Most scientists working with vertebrates use a classification based purely on phylogeny, organized by their known evolutionary history and sometimes disregarding the conventional interpretations of their anatomy and physiology.

Phylogenetic Relationships

In phylogenetic taxonomy, the relationships between animals are not typically divided into ranks, but illustrated as a nested "family tree" known as a cladogram. Phylogenetic groups are given definitions based on their relationship to one another, rather than purely on physical traits, such as the presence of a backbone. This nesting pattern is often combined with traditional taxonomy (as above), in a practice known as evolutionary taxonomy.

The cladogram presented below is based on studies compiled by Philippe Janvier and others for the *Tree of Life Web Project*.

Number of Extant Species

The number of described vertebrate species are split evenly between tetrapods and fish. The following table lists the number of described extant species for each vertebrate class as estimated in the IUCN Red List of Threatened Species, *2014.3*.

Vertebrate groups			Image	Class	Estimated number of described species	Group totals
Anamni-ote lack amniotic mem-brane so need to repro-duce in water	Jawless	Fish		Myxini (hagfish)		32,900
				Hyperoartia (lamprey)		
	Jawed			cartilaginous fish		
				ray-finned fish		
				lobe-finned fish		
		Tetrapods		amphibians	7,302	33,278
Amniote have amniotic mem-brane adapted to reproduc-ing on land				reptiles	10,038	
				birds	10,425	
				mammals	5,513	
				Total described species		**66,178**

The IUCN estimates that 1,305,075 extant invertebrate species have been described, which means that less than 5% of the described animal species in the world are vertebrates.

Reproductive Systems

Nearly all vertebrates undergo sexual reproduction. They produce haploid gametes by meiosis. The smaller, motile gametes are spermatozoa and the larger, non-motile gametes are ova. These fuse by the process of fertilisation to form diploid zygotes, which develop into new individuals.

Inbreeding

During sexual reproduction, mating with a close relative (inbreeding) often leads to inbreeding depression. Inbreeding depression is considered to be largely due to expression of deleterious recessive mutations. The effects of inbreeding have been studied in many vertebrate species.

In several species of fish, inbreeding was found to decrease reproductive success.

Inbreeding was observed to increase juvenile mortality in 11 small animal species.

A common breeding practice for pet dogs is mating between close relatives (e.g. between half- and full siblings). This practice generally has a negative effect on measures of reproductive success including decreased litter size and puppy survival.

Incestuous matings in birds result in severe fitness costs due to inbreeding depression (e.g. reduction in hatchability of eggs and reduced progeny survival).

Inbreeding Avoidance

As a result of the negative fitness consequences of inbreeding, vertebrate species have evolved mechanisms to avoid inbreeding. Numerous inbreeding avoidance mechanisms operating prior to mating have been described.

Toads display breeding site fidelity, as do many amphibians. Individuals that return to natal ponds to breed will likely encounter siblings as potential mates. Although incest is possible, *Bufo americanus* siblings rarely mate. These toads likely recognize and actively avoid close kins as mates. Advertisement vocalizations by males appear to serve as cues by which females recognize their kin.

Inbreeding avoidance mechanisms can also operate subsequent to copulation. In guppies, a post-copulatory mechanism of inbreeding avoidance occurs based on competition between sperm of rival males for achieving fertilization. In competitions between sperm from an unrelated male and from a full sibling male, a significant bias in paternity towards the unrelated male was observed.

When female sand lizards mate with two or more males, sperm competition within the female's reproductive tract may occur. Active selection of sperm by females appears to occur in a manner that enhances female fitness. On the basis of this selective process, the sperm of males that are more distantly related to the female are preferentially used for fertilization, rather than the sperm of close relatives. This preference may enhance the fitness of progeny by reducing inbreeding depression.

Outcrossing

Mating with unrelated or distantly related members of the same species is generally thought to provide the advantage of masking deleterious recessive mutations in progeny. Vertebrates have evolved numerous diverse mechanisms for avoiding close inbreeding and promoting outcrossing.

Outcrossing as a way of avoiding inbreeding depression, has been especially well studied in birds. For instance, inbreeding depression occurs in the great tit when the offspring are produced as a result of a mating between close relatives. In natural populations of the great tit (*Parus major*), inbreeding is avoided by dispersal of individuals from their birthplace which reduces the chance of mating with a close relative.

The purple-crowned fairywren females paired with related males may undertake extra-pair matings that can reduce the negative effects of inbreeding. However, there are ecological and demographic constraints on extra pair matings. Nevertheless, 46% of broods produced by incestuously paired females contained extra-pair young.

Southern pied babblers (*Turdoides bicolor*) appear to avoid inbreeding in two ways. The first is through dispersal, and the second is by avoiding familiar group members as mates. Although both males and females disperse locally, they move outside the range where genetically related individuals are likely to be encountered. Within their group, individuals only acquire breeding positions when the opposite-sex breeder is unrelated.

Cooperative breeding in birds typically occurs when offspring, usually males, delay dispersal from their natal group in order to remain with the family to help rear younger kin. Female offspring rarely stay at home, dispersing over distances that allow them to breed independently, or to join unrelated groups.

Parthenogenesis

Parthenogenesis is a natural form of reproduction in which growth and development of embryos occur without fertilization. *Agkistrodon contortrix* (copperhead snake) and *Agkistrodon piscivorus* (cotton mouth snake) can reproduce by facultative parthenogenesis. That is, they are capable of switching from a sexual mode of reproduction to an asexual mode.

Reproduction in squamate reptiles is ordinarily sexual, with males having a ZZ pair of sex determining chromosomes, and females a ZW pair. However, the Colombian Rainbow boa (*Epicrates maurus*) can also reproduce by facultative parthenogenesis resulting in production of WW female progeny. The WW females are likely produced by terminal automixis.

Mole salamanders are an ancient (2.4-3.8 million year-old) unisexual vertebrate lineage. In the polyploid unisexual mole salamander females, a premeiotic endomitotic event doubles the number of chromosomes. As a result, the mature eggs produced subsequent to the two meiotic divisions have the same ploidy as the somatic cells of the female salamander. Synapsis and recombination during meiotic prophase I in these unisexual females is thought to ordinarily occur between identical sister chromosomes and occasionally between homologous chromosomes. Thus little, if any, genetic variation is produced. Recombination between homeologous chromosomes occurs only rarely, if at all. Since production of genetic variation is weak, at best, it is unlikely to provide a benefit sufficient to account for the long-term maintenance of meiosis in these organisms. However, meiosis may have been maintained during evolution by the efficient recombinational repair of DNA damages that meiosis provides, an advantage that could be realized at each generation.

Self-fertilization

The mangrove killifish (*Kryptolebias marmoratus*) produces both eggs and sperm by meiosis and routinely reproduces by self-fertilisation. The capacity for selfing in these fishes has apparently persisted for at least several hundred thousand years. Each individual hermaphrodite normally fertilizes itself when an egg and sperm that it has produced by an internal organ unite inside the fish's body. In nature, this mode of reproduction can yield highly homozygous lines composed of individuals so genetically uniform as to be, in effect, identical to one another. Although inbreeding, especially in the extreme form of self-fertilization, is ordinarily regarded as detrimental because it leads to expression of deleterious recessive alleles, self-fertilization does provide the benefit of *fertilization assurance* (reproductive assurance) at each generation.

Marine Vertebrate

Marine vertebrates are vertebrates which live in a marine environment. These primarily include fish, seabirds, marine reptiles, and marine mammals. These animals have an internal skeleton and make up about 4% of the sea's animal population.

Invertebrate

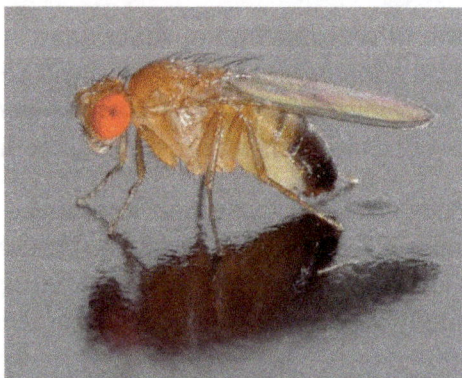

The common fruit fly, *Drosophila melanogaster*, has been used extensively for research

Invertebrates are animals that neither possess nor develop a vertebral column (commonly known as a backbone or *spine*), derived from the notochord. This includes all animals apart from the subphylum Vertebrata. Familiar examples of invertebrates include insects; crabs, lobsters and their kin; snails, clams, octopuses and their kin; starfish, sea-urchins and their kin; and worms. The majority of animal species are invertebrates; one estimate puts the figure at 97%. Many invertebrate taxa have a greater number and variety of species than the entire subphylum of Vertebrata.

Some of the so-called invertebrates, such as the Chaetognatha, Hemichordata, Tunicata and Cephalochordata are more closely related to the vertebrates than to other invertebrates. This makes the term "invertebrate" paraphyletic and hence almost meaningless for taxonomic purposes.

Etymology

The word "invertebrate" comes from the form of the Latin word *vertebra*, which means a joint in general, and sometimes specifically a joint from the spinal column of a vertebrate. In turn the jointed aspect of *vertebra* derived from the concept of turning, expressed in the root *verto* or *vorto*, to turn. Coupled with the prefix *in-*, meaning "not" or "without".

Taxonomic Significance

The term invertebrates is not always precise among non-biologists since it does not accurately describe a taxon in the same way that Arthropoda, Vertebrata or Manidae do. Each of these terms describes a valid taxon, phylum, subphylum or family. "Invertebrata" is a term of convenience, not a taxon; it has very little circumscriptional significance except within the Chordata. The Vertebrata as a subphylum comprises such a small proportion of the Metazoa that to speak of the kingdom Animalia in terms of "Vertebrata" and "Invertebrata" has limited practicality. In the more formal taxonomy of Animalia other attributes that logically should precede the presence or absence of the vertebral column in constructing a cladogram, for example, the presence of a notochord. That would at least circumscribe the Chordata. However, even the notochord would be a less fundamental criterion than aspects of embryological development and symmetry or perhaps bauplan.

Invertebrates don't have a skeleton of bone, either internal or external. They include hugely varied body plans. Many have fluid-filled, hydrostatic skeletons, like jellyfish or worms. Others have hard exoskeletons, outer shells like those of insects and crustaceans. The most familiar invertebrates include the Protozoa, Porifera, Coelenterata, Platyhelminthes, Nematoda, Annelida, Echinodermata, Mollusca and Arthropoda. Arthropoda include insects, crustaceans and arachnids.

Number of Extant Species

By far the largest number of described invertebrate species are insects. The following table lists the number of described extant species for major invertebrate groups as estimated in the IUCN Red List of Threatened Species, *2014.3*.

Invertebrate group	Image	Estimated number of described species
Insects		1,000,000
Molluscs		85,000

Crustaceans		47,000
Corals		2,000
Arachnids		102,248
Velvet worms		165
Horseshoe crabs		4
Others jellyfish, echinoderms, sponges, other worms etc.		68,658
	Totals:	**1,305,075**

The IUCN estimates that 66,178 extant vertebrate species have been described, which means that over 95% of the described animal species in the world are invertebrates.

Characteristics

The trait that is common to all invertebrates is the absence of a vertebral column (backbone): this creates a distinction between invertebrates and vertebrates. The distinction is one of convenience only; it is not based on any clear biologically homologous trait, any more than the common trait of having wings functionally unites insects, bats, and birds, or than not having wings unites tortoises, snails and sponges. Being animals, invertebrates are heterotrophs, and require sustenance in the form of the consumption of other organisms. With a few exceptions, such as the Porifera, invertebrates generally have bodies composed of differentiated tissues. There is also typically a digestive chamber with one or two openings to the exterior.

Morphology and Symmetry

The body plans of most multicellular organisms exhibit some form of symmetry, whether radial, bilateral, or spherical. A minority, however, exhibit no symmetry. One example of asymmetric invertebrates include all gastropod species. This is easily seen in snails and sea snails, which have helical shells. Slugs appear externally symmetrical, but their pneumostome (breathing hole) is located on the right side. Other gastropods develop external asymmetry, such as Glaucus atlanticus that develops asymmetrical cerata as they mature. The origin of gastropod asymmetry is a subject of scientific debate.

Other examples of asymmetry are found in fiddler crabs and hermit crabs. They often have one claw much larger than the other. If a male fiddler loses its large claw, it will grow another on the opposite side after moulting. Sessile animals such as sponges are asymmetrical alongside coral colonies (with the exception of the individual polyps that exhibit radial symmetry); alpheidae claws that lack pincers; and some copepods, polyopisthocotyleans, and monogeneans which parasitize by attachment or residency within the gill chamber of their fish hosts).

Nervous System

Neurons differ in invertebrates from mammalian cells. Invertebrates cells fire in response to similar stimuli as mammals, such as tissue trauma, high temperature, or changes in pH. The first invertebrate in which a neuron cell was identified was the medicinal leech, *Hirudo medicinalis*.

Learning and memory using nociceptors in the sea hare, *Aplysia* has been described. Mollusk neurons are able to detect increasing pressures and tissue trauma.

Neurons have been identified in a wide range of invertebrate species, including annelids, molluscs, nematodes and arthropods.

Respiratory System

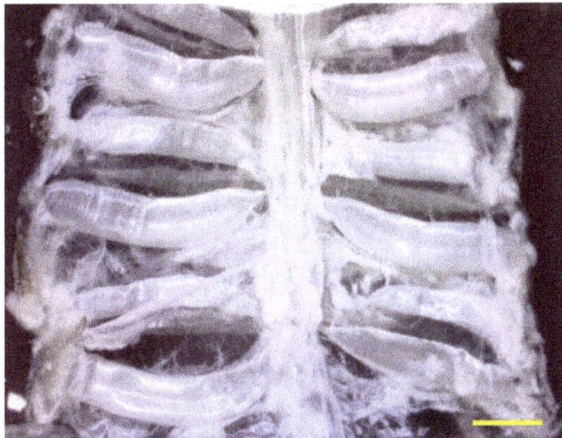

Tracheal system of dissected cockroach. The largest tracheae run across the width of the body of the cockroach and are horizontal in this image. Scale bar, 2 mm.

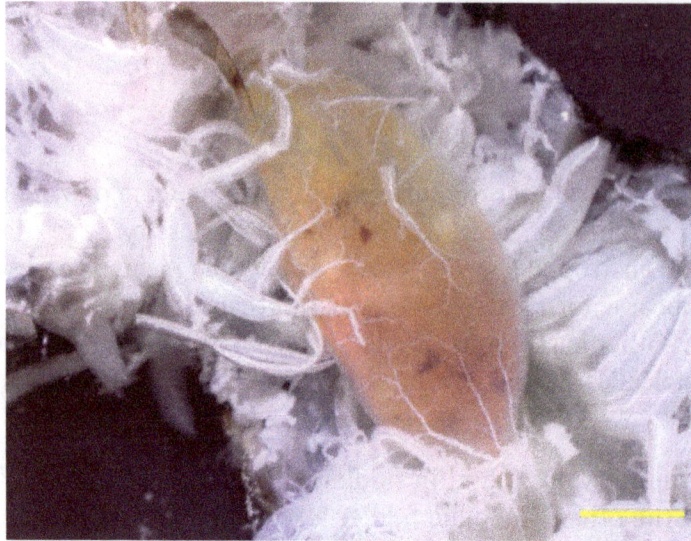

The tracheal system branches into progressively smaller tubes, here supplying the crop of the cockroach.
Scale bar, 2.0 mm.

One type of invertebrate respiratory system is the open respiratory system composed of spiracles, tracheae, and tracheoles that terrestrial arthropods have to transport metabolic gases to and from tissues. The distribution of spiracles can vary greatly among the many orders of insects, but in general each segment of the body can have only one pair of spiracles, each of which connects to an atrium and has a relatively large tracheal tube behind it. The tracheae are invaginations of the cuticular exoskeleton that branch (anastomose) throughout the body with diameters from only a few micrometres up to 0.8 mm. The smallest tubes, tracheoles, penetrate cells and serve as sites of diffusion for water, oxygen, and carbon dioxide. Gas may be conducted through the respiratory system by means of active ventilation or passive diffusion. Unlike vertebrates, insects do not generally carry oxygen in their haemolymph.

A tracheal tube may contain ridge-like circumferential rings of taenidia in various geometries such as loops or helices. In the head, thorax, or abdomen, tracheae may also be connected to air sacs. Many insects, such as grasshoppers and bees, which actively pump the air sacs in their abdomen, are able to control the flow of air through their body. In some aquatic insects, the tracheae exchange gas through the body wall directly, in the form of a gill, or function essentially as normal, via a plastron. Note that despite being internal, the tracheae of arthropods are shed during moulting (ecdysis).

Digestive System

Integumentary System

Reproduction

Like vertebrates, most invertebrates reproduce at least partly through sexual reproduction. They produce specialized reproductive cells that undergo meiosis to produce smaller, motile spermatozoa or larger, non-motile ova. These fuse to form zygotes, which develop into new individuals. Others are capable of asexual reproduction, or sometimes, both methods of reproduction.

Social Interaction

Social behavior is widespread in invertebrates, including cockroaches, termites, aphids, thrips, ants, bees, Passalidae, Acari, spiders, and more. Social interaction is particularly salient in eusocial species but applies to other invertebrates as well.

Insects recognize information transmitted by other insects.

Phyla

The fossil coral *Cladocora* from the Pliocene of Cyprus

The term invertebrates covers several phyla. One of these are the sponges (Porifera). They were long thought to have diverged from other animals early. They lack the complex organization found in most other phyla. Their cells are differentiated, but in most cases not organized into distinct tissues. Sponges typically feed by drawing in water through pores. Some speculate that sponges are not so primitive, but may instead be secondarily simplified. The Ctenophora and the Cnidaria, which includes sea anemones, corals, and jellyfish, are radially symmetric and have digestive chambers with a single opening, which serves as both the mouth and the anus. Both have distinct tissues, but they are not organized into organs. There are only two main germ layers, the ectoderm and endoderm, with only scattered cells between them. As such, they are sometimes called diploblastic.

The Echinodermata are radially symmetric and exclusively marine, including starfish (Asteroidea), sea urchins, (Echinoidea), brittle stars (Ophiuroidea), sea cucumbers (Holothuroidea) and feather stars (Crinoidea).

The largest animal phylum is also included within invertebrates: the Arthropoda, including insects, spiders, crabs, and their kin. All these organisms have a body divided into repeating segments, typically with paired appendages. In addition, they possess a hardened exoskeleton that is periodically shed during growth. Two smaller phyla, the Onychophora and Tardigrada, are close relatives of the arthropods and share these traits. The Nematoda or roundworms, are perhaps the second largest animal phylum, and are also invertebrates. Roundworms are typically microscopic, and occur in nearly every environment where there is water. A number are important parasites. Smaller phyla related to them are the Kinorhyncha, Priapulida, and Loricifera. These groups have a reduced coelom, called a pseudocoelom. Other invertebrates include the Nemertea or ribbon worms, and the Sipuncula.

Another phylum is Platyhelminthes, the flatworms. These were originally considered primitive, but it now appears they developed from more complex ancestors. Flatworms are acoelomates, lacking a body cavity, as are their closest relatives, the microscopic Gastrotricha. The Rotifera or rotifers, are common in aqueous environments. Invertebrates also include the Acanthocephala or spiny-headed worms, the Gnathostomulida, Micrognathozoa, and the Cycliophora.

Also included are two of the most successful animal phyla, the Mollusca and Annelida. The former, which is the second-largest animal phylum by number of described species, includes animals such as snails, clams, and squids, and the latter comprises the segmented worms, such as earthworms and leeches. These two groups have long been considered close relatives because of the common presence of trochophore larvae, but the annelids were considered closer to the arthropods because they are both segmented. Now, this is generally considered convergent evolution, owing to many morphological and genetic differences between the two phyla.

Among lesser phyla of invertebrates are the Hemichordata, or acorn worms, and the Chaetognatha, or arrow worms. Other phyla include Acoelomorpha, Brachiopoda, Bryozoa, Entoprocta, Phoronida, and Xenoturbellida.

Classification of Invertebrates

Invertebrates can be classified into several main categories, some of which are taxonomically obsolescent or debatable, but still used as terms of convenience. Each however appears in its own article at the following links.

- *Protozoa* (like the worms, an arbitrary grouping of convenience)
- Sponges (*Porifera*)
- Stinging jellyfish and corals (*Cnidaria*)
- Comb jellies (*Ctenophora*)
- Flatworms (*Platyhelminthes*)
- Round- or threadworms (*Nematoda*)
- segmented worms (*Annelida*)
- Insects, spiders, crabs and their kin (*Arthropoda*)
- Cuttlefish, snails, mussels and their kin (*Mollusca*)
- Starfish, sea-cucumbers and their kin (*Echinodermata*)

History

The earliest animal fossils appear to be those of invertebrates. 665-million-year-old fossils in the Trezona Formation at Trezona Bore, West Central Flinders, South Australia have been interpreted as being early sponges. Some paleontologists suggest that animals appeared much earlier, possibly as early as 1 billion years ago. Trace fossils such as tracks and burrows found in the Tonian era indicate the presence of triploblastic worms, like metazoans, roughly as large (about 5 mm wide) and complex as earthworms.

Around 453 MYA, animals began diversifying, and many of the important groups of invertebrates diverged from one another. Fossils of invertebrates are found in various types of sediment from the Phanerozoic. Fossils of invertebrates are commonly used in stratigraphy.

Classification

Carl Linnaeus divided these animals into only two groups, the Insecta and the now-obsolete Vermes (worms). Jean-Baptiste Lamarck, who was appointed to the position of "Curator of Insecta and Vermes" at the Muséum National d'Histoire Naturelle in 1793, both coined the term "invertebrate" to describe such animals, and divided the original two groups into ten, by splitting Arachnida and Crustacea from the Linnean Insecta, and Mollusca, Annelida, Cirripedia, Radiata, Coelenterata and Infusoria from the Linnean Vermes. They are now classified into over 30 phyla, from simple organisms such as sea sponges and flatworms to complex animals such as arthropods and molluscs.

Significance of the Group

Invertebrates are animals *without* a vertebral column. This has led to the conclusion that *inverte-brates* are a group that deviates from the normal, vertebrates. This has been said to be because researchers in the past, such as Lamarck, viewed vertebrates as a "standard": in Lamarck's theory of evolution, he believed that characteristics acquired through the evolutionary process involved not only survival, but also progression toward a "higher form", to which humans and vertebrates were closer than invertebrates were. Although goal-directed evolution has been abandoned, the distinction of invertebrates and vertebrates persists to this day, even though the grouping has been noted to be "hardly natural or even very sharp." Another reason cited for this continued distinction is that Lamarck created a precedent through his classifications which is now difficult to escape from. It is also possible that some humans believe that, they themselves being vertebrates, the group deserves more attention than invertebrates. In any event, in the 1968 edition of *Invertebrate Zoology*, it is noted that "division of the Animal Kingdom into vertebrates and invertebrates is artificial and reflects human bias in favor of man's own relatives." The book also points out that the group lumps a vast number of species together, so that no one characteristic describes all invertebrates. In addition, some species included are only remotely related to one another, with some more related to vertebrates than other invertebrates.

In Research

For many centuries, invertebrates have been neglected by biologists, in favor of big vertebrates and "useful" or charismatic species. Invertebrate biology was not a major field of study until the work of Linnaeus and Lamarck in the 18th century. During the 20th century, invertebrate zoology became one of the major fields of natural sciences, with prominent discoveries in the fields of medicine, genetics, palaeontology, and ecology. The study of invertebrates has also benefited law enforcement, as arthropods, and especially insects, were discovered to be a source of information for forensic investigators.

Two of the most commonly studied model organisms nowadays are invertebrates: the fruit fly *Drosophila melanogaster* and the nematode *Caenorhabditis elegans*. They have long been the most intensively studied model organisms, and were among the first life-forms to be genetically sequenced. This was facilitated by the severely reduced state of their genomes, but many genes, introns, and linkages have been lost. Analysis of the starlet sea anemone genome has emphasised the importance of sponges, placozoans, and choanoflagellates, also being sequenced, in explaining the arrival of 1500 ancestral genes unique to animals. Invertebrates are also used by scientists in the field of aquatic biomonitoring to evaluate the effects of water pollution and climate change.

Marine Invertebrates

Ernst Haeckel's 96th plate, showing some marine invertebrates. Marine invertebrates have a large variety of body plans, which are currently categorised into over 30 phyla.

Marine invertebrates are the invertebrates that live in marine habitats. Invertebrate is a blanket term that includes all animals apart from the vertebrate members of the chordate phylum. Invertebrates lack a vertebral column, and some have evolved a shell or a hard exoskeleton. As on land and in the air, marine invertebrates have a large variety of body plans, and have been categorised into over 30 phyla. They make up most of the macroscopic life in the oceans.

Evolution

Kimberella, an early mollusc important for understanding the Cambrian explosion. Invertebrates are grouped into different phyla (body plans).

Opabinia an extinct, stem group arthropod that appeared in the Middle Cambrian

The earliest animals were marine invertebrates, that is, vertebrates came later. Animals are multicellular eukaryotes, and are distinguished from plants, algae, and fungi by lacking cell walls. Marine invertebrates are animals that inhabit a marine environment apart from the vertebrate members of the chordate phylum; invertebrates lack a vertebral column. Some have evolved a shell or a hard exoskeleton.

The earliest widely accepted animal fossils are the rather modern-looking cnidarians (the group that includes jellyfish, sea anemones and *Hydra*), possibly from around 580 Ma The Ediacara biota, which flourished for the last 40 million years before the start of the Cambrian, were the first animals more than a very few centimetres long. Many were flat and had a "quilted" appearance, and seemed so strange that there was a proposal to classify them as a separate kingdom, Vendozoa. Others, however, have been interpreted as early molluscs (*Kimberella*), echinoderms (*Arkarua*), and arthropods (*Spriggina*, *Parvancorina*). There is still debate about the classification of these specimens, mainly because the diagnostic features which allow taxonomists to classify more recent organisms, such as similarities to living organisms, are generally absent in the Ediacarans. However, there seems little doubt that *Kimberella* was at least a triploblastic bilaterian animal, in other words, an animal significantly more complex than the cnidarians.

The small shelly fauna are a very mixed collection of fossils found between the Late Ediacaran and Middle Cambrian periods. The earliest, *Cloudina*, shows signs of successful defense against predation and may indicate the start of an evolutionary arms race. Some tiny Early Cambrian shells almost certainly belonged to molluscs, while the owners of some "armor plates," *Halkieria* and *Microdictyon*, were eventually identified when more complete specimens were found in Cambrian lagerstätten that preserved soft-bodied animals.

In the 1970s there was already a debate about whether the emergence of the modern phyla was "explosive" or gradual but hidden by the shortage of Precambrian animal fossils. A re-analysis of fossils from the Burgess Shale lagerstätte increased interest in the issue when it revealed animals, such as *Opabinia*, which did not fit into any known phylum. At the time these were interpreted as evidence that the modern phyla had evolved very rapidly in the Cambrian explosion and that the Burgess Shale's "weird wonders" showed that the Early Cambrian was a uniquely experimental period of animal evolution. Later discoveries of similar animals and the development of new theoretical approaches led to the conclusion that many of the "weird wonders" were evolutionary "aunts" or "cousins" of modern groups—for example that *Opabinia* was a member of the lobopods, a group which includes the ancestors of the arthropods, and that it may have been closely related to the modern tardigrades. Nevertheless, there is still much debate about whether the Cambrian explosion was really explosive and, if so, how and why it happened and why it appears unique in the history of animals.

Classification

Invertebrates are grouped into different phyla. Informally phyla can be thought of as a way of grouping organisms according to their body plan. A body plan refers to a blueprint which describes the shape or morphology of an organism, such as its symmetry, segmentation and the disposition of its appendages. The idea of body plans originated with vertebrates, which were grouped into one phylum. But the vertebrate body plan is only one of many, and invertebrates consist of many phyla or body plans. The history of the discovery of body plans can be seen as a movement from a worldview centred on vertebrates, to seeing the vertebrates as one body plan among many. Among the pioneering zoologists, Linnaeus identified two body plans outside the vertebrates; Cuvier identified three; and Haeckel had four, as well as the Protista with eight more, for a total of twelve. For comparison, the number of phyla recognised by modern zoologists has risen to 35.

Historically body plans were thought of as having evolved in rapidly during the Cambrian explosion, but a more nuanced understanding of animal evolution suggests a gradual development of body plans throughout the early Palaeozoic and beyond. More generally a phylum can be defined in two ways: as described above, as a group of organisms with a certain degree of morphological or developmental similarity (the phenetic definition), or a group of organisms with a certain degree of evolutionary relatedness (the phylogenetic definition).

As on land and in the air, invertebrates make up a great majority of all macroscopic life, as the vertebrates makes up a subphylum of one of over 30 known animal phyla, making the term almost meaningless for taxonomic purpose. Invertebrate sea life includes the following groups, some of which are phyla:

The 49th plate from Ernst Haeckel's *Kunstformen der Natur*, 1904, showing various sea anemones classified as Actiniae, in the Cnidaria phylum

"A variety of marine worms": plate from *Das Meer* by M.J. Schleiden (1804–1881)

- Acoela, among the most primitive bilateral animals;

- Annelida, (polychaetes and sea leeches);

- Brachiopoda, marine animals that have hard "valves" (shells) on the upper and lower surfaces ;

- Bryozoa, also known as moss animals or sea mats;

- Chaetognatha, commonly known as arrow worms, are a phylum of predatory marine worms that are a major component of plankton;

- Cephalochordata represented in the modern oceans by the lancelets (also known as Amphioxus);

- Cnidaria, such as jellyfish, sea anemones, and corals;

- Crustacea, including lobsters, crabs, shrimp, crayfish, barnacles, hermit crabs, mantis shrimps, and copepods;

- Ctenophora, also known as comb jellies, the largest animals that swim by means of cilia;

- Echinodermata, including sea stars, brittle stars, sea urchins, sand dollars, sea cucumbers, crinoids, and sea daisies;

- Echiura, also known as spoon worms;

- Gnathostomulids, slender to thread-like worms, with a transparent body that inhabit sand and mud beneath shallow coastal waters;

- Gastrotricha, often called hairy backs, found mostly interstitially in between sediment particles;

- Hemichordata, includes acorn worms, solitary worm-shaped organisms;

- Kamptozoa, goblet-shaped sessile aquatic animals, with relatively long stalks and a "crown" of solid tentacles, also called Entoprocta;

- Kinorhyncha, segmented, limbless animals, widespread in mud or sand at all depths, also called mud dragons;

- Loricifera, very small to microscopic marine sediment-dwelling animals only discovered in 1983;

- Merostomata; also known as horseshoe crabs;

- Mollusca, including shellfish, squid, octopus, whelks, *Nautilus*, cuttlefish, nudibranchs, scallops, sea snails, Aplacophora, Caudofoveata, Monoplacophora, Polyplacophora, and Scaphopoda;

- Myzostomida, a taxonomic group of small marine worms which are parasitic on crinoids or "sea lilies";

- Nemertinea, also known as "ribbon worms" or "proboscis worms";

- Orthonectida, a small phylum of poorly known parasites of marine invertebrates that are among the simplest of multi-cellular organisms;

- Phoronida, a phylum of marine animals that filter-feed with a lophophore (a "crown" of tentacles), and build upright tubes of chitin to support and protect their soft bodies;

- Placozoa, small, flattened, multicellular animals around 1 millimetre across and the simplest in structure. They have no regular outline, although the lower surface is somewhat concave, and the upper surface is always flattened;

- Porifera (sponges), multicellular organisms that have bodies full of pores and channels allowing water to circulate through them;

- Priapulida, or penis worms, are a phylum of marine worms that live marine mud. They are named for their extensible spiny proboscis, which, in some species, may have a shape like that of a human penis;

- Pycnogonida, also called sea spiders, are unrelated to spiders, or even to arachnids which they resemble;

- Sipunculida, also called peanut worms, is a group containing 144–320 species (estimates vary) of bilaterally symmetrical, unsegmented marine worms;

- Tunicata, also known as sea squirts or sea pork, are filter feeders attached to rocks or similarly suitable surfaces on the ocean floor;

- Some flatworms of the classes Turbellaria and Monogenea;

- Xenoturbella, a genus of bilaterian animals that contains only two marine worm-like species;

- Xiphosura, includes a large number of extinct lineages and only four recent species in the family Limulidae, which include the horseshoe crabs.

Arthropods total about 1,113,000 described extant species, molluscs about 85,000 and chordates about 52,000.

Marine Sponges

Sponges have no nervous, digestive or circulatory system

Sponges are animals of the phylum Porifera (Modern Latin for *bearing pores*). They are multi-cellular organisms that have bodies full of pores and channels allowing water to circulate through them, consisting of jelly-like mesohyl sandwiched between two thin layers of cells. They have unspecialized cells that can transform into other types and that often migrate between the main cell layers and the mesohyl in the process. Sponges do not have nervous, digestive or circulatory systems. Instead, most rely on maintaining a constant water flow through their bodies to obtain food and oxygen and to remove wastes.

Sponges are similar to other animals in that they are multicellular, heterotrophic, lack cell walls and produce sperm cells. Unlike other animals, they lack true tissues and organs, and have no body symmetry. The shapes of their bodies are adapted for maximal efficiency of water flow through the central cavity, where it deposits nutrients, and leaves through a hole called the osculum. Many sponges have internal skeletons of spongin and/or spicules of calcium carbonate or silicon dioxide. All sponges are sessile aquatic animals. Although there are freshwater species, the great majority are marine (salt water) species, ranging from tidal zones to depths exceeding 8,800 m (5.5 mi).

While most of the approximately 5,000–10,000 known species feed on bacteria and other food particles in the water, some host photosynthesizing micro-organisms as endosymbionts and these alliances often produce more food and oxygen than they consume. A few species of sponge that live in food-poor environments have become carnivores that prey mainly on small crustaceans.

Sponge biodiversity. There are four sponge species in this photo.

Stove-pipe sponge

Linnaeus mistakenly identified sponges as plants in the order Algae. For a long time thereafter sponges were assigned to a separate subkingdom, Parazoa (meaning *beside the animals*). They are now classified as a paraphyletic phylum from which the higher animals have evolved.

Marine Cnidarians

Cnidarians are the simplest animals with cells organised into tissues. Yet the starlet sea anemone contains the same genes as those that form the vertebrate head.

Cnidarians (Greek for *nettle*) are distinguished by the presence of stinging cells, specialized cells that they use mainly for capturing prey. Cnidarians include corals, sea anemones, jellyfish and hydrozoans. They form a phylum containing over 10,000 species of animals found exclusively in aquatic (mainly marine) environments. Their bodies consist of mesoglea, a non-living jelly-like substance, sandwiched between two layers of epithelium that are mostly one cell thick. They have two basic body forms: swimming medusae and sessile polyps, both of which are radially symmetrical with mouths surrounded by tentacles that bear cnidocytes. Both forms have a single orifice and body cavity that are used for digestion and respiration.

Fossil cnidarians have been found in rocks formed about 580 million years ago. Fossils of cnidarians that do not build mineralized structures are rare. Scientists currently think cnidarians, ctenophores and bilaterians are more closely related to calcareous sponges than these are to other sponges, and that anthozoans are the evolutionary "aunts" or "sisters" of other cnidarians, and the most closely related to bilaterians.

Cnidarians are the simplest animals in which the cells are organised into tissues. The starlet sea anemone is used as a model organism in research. It is easy to care for in the laboratory and a protocol has been developed which can yield large numbers of embryos on a daily basis. There is a remarkable degree of similarity in the gene sequence conservation and complexity between the sea anemone and vertebrates. In particular, genes concerned in the formation of the head in vertebrates are also present in the anemone.

Moon jellyfish, found in coastal waters around the world

Lion's mane jellyfish, largest known jellyfish

Marine Worms

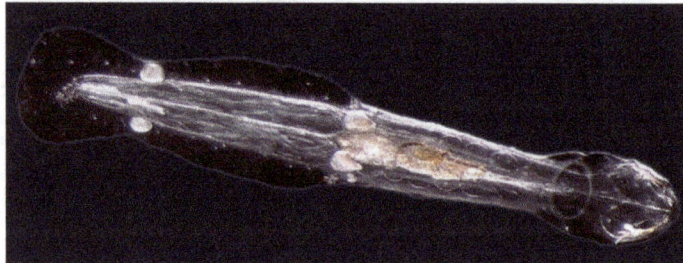

Arrow worms are predatory components of plankton worldwide.

Worms (Old English for *serpent*) typically have long cylindrical tube-like bodies and no limbs. Marine worms vary in size from microscopic to over 1 metre (3.3 ft) in length for some marine polychaete worms (bristle worms) and up to 58 metres (190 ft) for the marine nemertean worm (bootlace worm). Some marine worms occupy a small variety of parasitic niches, living inside the bodies of other animals, while others live more freely in the marine environment or by burrowing underground.

Different groups of marine worms are related only distantly, so they are found in several different phyla such as the Annelida (segmented worms), Chaetognatha (arrow worms), Hemichordata, and Phoronida (horseshoe worms). Many of these worms have specialized tentacles used for exchanging oxygen and carbon dioxide and also may be used for reproduction. Some marine worms are tube worms, such as the giant tube worm which lives in waters near underwater volcanoes and can withstand temperatures up to 90 degrees Celsius.

Platyhelminthes (flatworms) form another worm phylum which includes a class Cestoda of parasitic tapeworms. The marine tapeworm *Polygonoporus giganticus*, found in the gut of sperm whales, can grow to over 30 m (100 ft).

Nematodes (roundworms) constitute a further worm phylum with tubular digestive systems and an opening at both ends. Over 25,000 nematode species have been described, of which more than half are parasitic, t, and it has been estimated another million remain undescribed. They are ubiquitous in marine, freshwater and terrestrial environments, where they often outnumber other animals in both individual and species counts. They are found in every part of the earth's lithosphere, from the top of mountains to the bottom of oceanic trenches. By count they represent 90% of all animals on the ocean floor. Their numerical dominance, often exceeding a million individuals per square meter and accounting for about 80% of all individual animals on earth, their diversity of life cycles, and their presence at various trophic levels point at an important role in many ecosystems.

Bloodworms are typically found on the bottom of shallow marine waters

Echinoderms

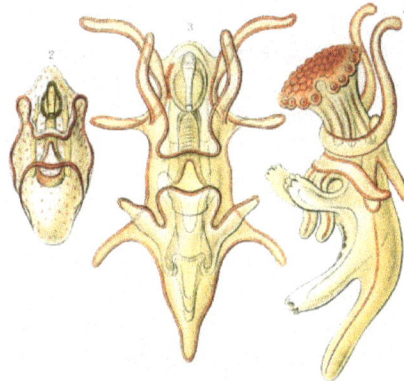

Starfish larvae are bilaterally symmetric, whereas the adults have fivefold symmetry

Echinoderms (Greek for *spiny skin*) is a phylum which contains only marine invertebrates. The adults are recognizable by their radial symmetry (usually five-point) and include starfish, sea urchins, sand dollars, and sea cucumbers, as well as the sea lilies. Echinoderms are found at every ocean depth, from the intertidal zone to the abyssal zone. The phylum contains about 7000 living species, making it the second-largest grouping of deuterostomes (a superphylum), after the chordates (which include the vertebrates, such as birds, fishes, mammals, and reptiles).

Echinoderms are unique among animals in having bilateral symmetry at the larval stage, but fivefold symmetry (pentamerism, a special type of radial symmetry) as adults.

The echinoderms are important both biologically and geologically. Biologically, there are few other groupings so abundant in the biotic desert of the deep sea, as well as shallower oceans. The more notably distinct trait, which most echinoderms have, is their remarkable powers of regeneration of tissue, organs, limbs, and of asexual reproduction, and in some cases, complete regeneration from a single limb. Geologically, the value of echinoderms is in their ossified skeletons, which are major contributors to many limestone formations, and can provide valuable clues as to the geological environment. They were the most used species in regenerative research in the 19th and 20th centuries. Further, it is held by some scientists that the radiation of echinoderms was responsible for the Mesozoic Marine Revolution.

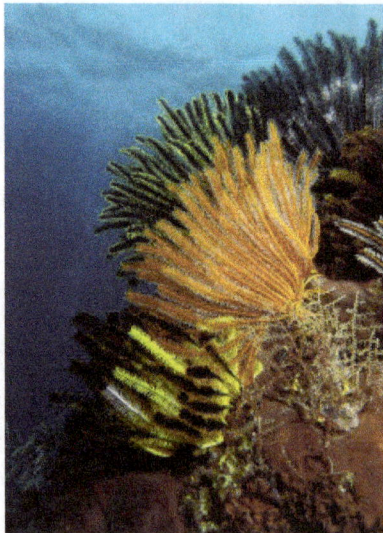

Colorful sea lilies in shallow waters

Aside from the hard-to-classify *Arkarua* (a Precambrian animal with echinoderm-like pentamerous radial symmetry), the first definitive members of the phylum appeared near the start of the Cambrian.

Marine Molluscs

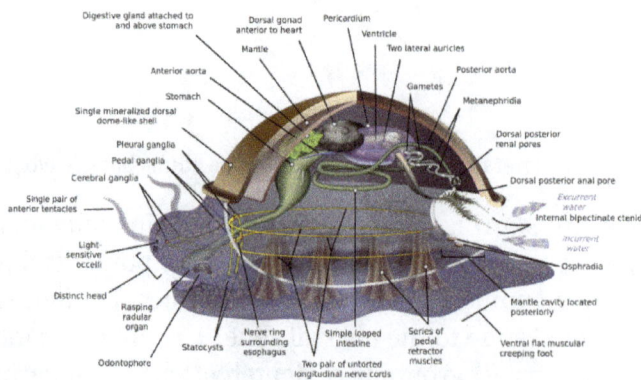

Generalized or *hypothetical ancestral mollusc*

Molluscs (Latin for *soft*) form a phylum with about 85,000 extant recognized species. They are the largest marine phylum, comprising about 23% of all the named marine organisms. Molluscs have more varied forms than other animal phylums. They are highly diverse, not just in size and in anatomical structure, but also in behaviour and in habitat. The majority of species still live in the oceans, from the seashores to the abyssal zone, but some form a significant part of the freshwater fauna and the terrestrial ecosystems. The phylum is divided into 9 or 10 taxonomic classes, two of which are extinct. They include snails, slugs and other gastropods; clams and other bivalves; squids and other cephalopods; and other lesser-known but similarly distinctive subgroups. Gastropods (snails and slugs) are by far the most numerous molluscs in terms of classified species, and account for 80% of the total. Cephalopoda such as squid, cuttlefish, and octopuses are among the neurologically most advanced of all invertebrates.

Molluscs have such diverse shapes that many textbooks base their descriptions of molluscan anatomy on a generalized or *hypothetical ancestral mollusc*. This generalized mollusc is unsegmented and bilaterally symmetrical with an underside consisting of a single muscular foot. Beyond that it has three further key features. Firstly, it has a muscular cloak called a mantle covering its viscera and containing a significant cavity used for breathing and excretion. A shell secreted by the mantle covers the upper surface. Secondly (apart from bivalves) it has a rasping tongue called a radula used for feeding. Thirdly, it has a nervous system including a complex digestive system using microscopic, muscle-powered hairs called cilia to exude mucus. The generalized mollusc has two paired nerve cords (three in bivalves). The brain, in species that have one, encircles the esophagus. Most molluscs have eyes and all have sensors detecting chemicals, vibrations, and touch. The simplest type of molluscan reproductive system relies on external fertilization, but more complex variations occur. All produce eggs, from which may emerge trochophore larvae, more complex veliger larvae, or miniature adults. The depiction is rather similar to modern monoplacophorans, and some suggest it may resemble very early molluscs.

In 1969 David Nicol estimated the probable total number of living molluscs at 107,000 of which about were 80,000 marine. In 2009, Chapman estimated the number of described living species at 85,000. Haszprunar in 2001 estimated about 93,000 named species, which include 23% of all named marine organisms. About 200,000 living species in total are estimated, and 70,000 fossil species, although the total number of mollusc species ever to have existed, whether or not preserved, must be many times greater than the number alive today.

Molluscs usually have eyes. Bordering the edge of the mantle of a scallop, a bivalve mollusc, can be over 100 simple eyes.

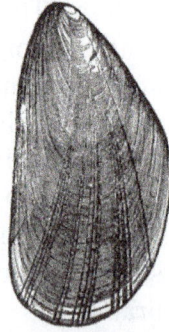

Common mussel, another bivalve

Good evidence exists for the appearance of marine gastropods, cephalopods and bivalves in the Cambrian period 541 to 485.4 million years ago. However, the evolutionary history both of molluscs' emergence from the ancestral Lophotrochozoa and of their diversification into the well-known living and fossil forms are still subjects of vigorous debate among scientists.

Marine Arthropods

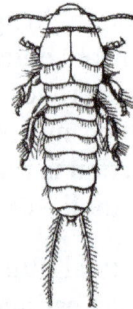

Hemimerus Hanseni.
(After Hansen.)

Head
_____
Thorax
_____
Abdomen
_____

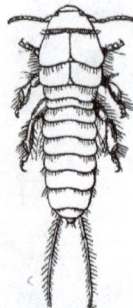

Hemimerus Hanseni.
(After Hansen.)

Segmentation and tagmata of an arthropod

Arthropods (Greek for *jointed feet*) have an exoskeleton (external skeleton), a segmented body, and jointed appendages (paired appendages). They form a phylum which includes insects, arachnids, myriapods, and crustaceans. Arthropods are characterized by their jointed limbs and cuticle made of chitin, often mineralised with calcium carbonate. The arthropod body plan consists of segments, each with a pair of appendages. The rigid cuticle inhibits growth, so arthropods replace it periodically by moulting. Their versatility has enabled them to become the most species-rich members of all ecological guilds in most environments.

Marine arthropods range in size from the microscopic crustacean *Stygotantulus* to the Japanese spider crab. Arthropods' primary internal cavity is a hemocoel, which accommodates their internal organs, and through which their haemolymph - analogue of blood - circulates; they have open circulatory systems. Like their exteriors, the internal organs of arthropods are generally built of repeated segments. Their nervous system is "ladder-like", with paired ventral nerve cords running through all segments and forming paired ganglia in each segment. Their heads are formed by fusion of varying numbers of segments, and their brains are formed by fusion of the ganglia of these segments and encircle the esophagus. The respiratory and excretory systems of arthropods vary, depending as much on their environment as on the subphylum to which they belong.

Their vision relies on various combinations of compound eyes and pigment-pit ocelli: in most species the ocelli can only detect the direction from which light is coming, and the compound eyes are the main source of information, but the main eyes of spiders are ocelli that can form images and, in a few cases, can swivel to track prey. Arthropods also have a wide range of chemical and mechanical sensors, mostly based on modifications of the many setae (bristles) that project through their cuticles. Arthropods' methods of reproduction and development are diverse; all terrestrial species use internal fertilization, but this is often by indirect transfer of the sperm via an appendage or the ground, rather than by direct injection. Marine species all lay eggs and use either internal or external fertilization. Arthropod hatchlings vary from miniature adults to grubs that lack jointed limbs and eventually undergo a total metamorphosis to produce the adult form.

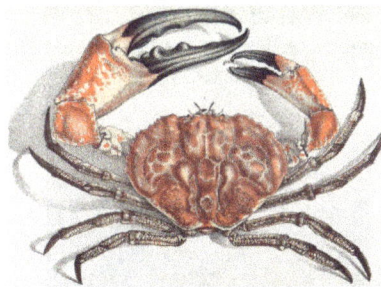

The Tasmanian giant crab is long-lived and slow-growing, making it vulnerable to overfishing.

The Japanese spider crab has the longest leg span of any arthropod.

The evolutionary ancestry of arthropods dates back to the Cambrian period. The group is generally regarded as monophyletic, and many analyses support the placement of arthropods with cyclo-neuralians (or their constituent clades) in a superphylum Ecdysozoa. Overall however, the basal relationships of Metazoa are not yet well resolved. Likewise, the relationships between various arthropod groups are still actively debated.

Other Phyla

Tardigrade, Lobopodia, (Onychophora)

- Non-craniate (non-vertebrate) chordates: Cephalochordate, Tunicata and *Haikouella*. These invertebrates are close relatives of the vertebrates.

- Non-craniate chordates are close relatives of vertebrates

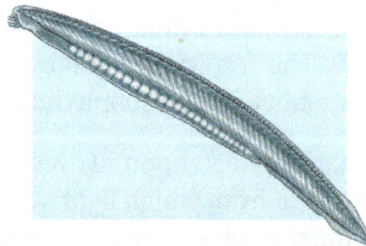

The lancelet, a small translucent fish-like Cephalochordate, is the closest living invertebrate relative of the vertebrates.

Fluorescent-colored sea squirts, *Rhopalaea crassa*. Tunicates may provide clues to vertebrate (and therefore human) ancestry.

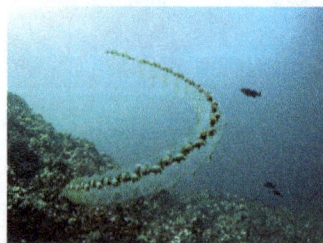

Salp chain

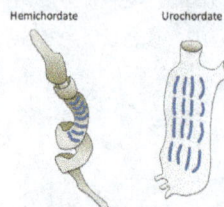

Gill slits in an acorn worm (left) and tunicate (right)

Minerals from Sea Water

There are a number of marine invertebrates that use minerals that are present in the sea in such minute quantities that they were undetectable until the advent of spectroscopy. Vanadium is concentrated by some tunicates for use in their blood cells to a level ten million times that of the surrounding seawater. Other tunicates similarly concentrate niobium and tantalum. Lobsters use copper in their respiratory pigment hemocyanin, despite the proportion of this metal in seawater being minute. Although these elements are present in vast quantities in the ocean, their extraction by man is not economic.

References

- Kardong, Kenneth V. (2002). Vertebrates: comparative anatomy, function, evolution. McGraw-Hill. pp. 288–289. ISBN 0-07-290956-0.

- Kardong, Kenneth V. (2002). Vertebrates: comparative anatomy, function, evolution. McGraw-Hill. pp. 287–288. ISBN 0-07-290956-0.

- Liem, Karel F.; Warren Franklin Walker (2001). Functional anatomy of the vertebrates: an evolutionary perspective. Harcourt College Publishers. p. 277. ISBN 978-0-03-022369-3.

- Benton, Michael J. (1 November 2004). Vertebrate Palaeontology (Third ed.). Blackwell Publishing. pp. 33, 455 pp. ISBN 978-0632056378.

- Wasserthal, Lutz T. (1998). Chapter 25: The Open Hemolymph System of Holometabola and Its Relation to the Tracheal Space. In "Microscopic Anatomy of Invertebrates". Wiley-Liss, Inc. ISBN 0-471-15955-7.

- Bhamrah, H. S.; Kavita Juneja (2003). An Introduction to Porifera. Anmol Publications PVT. LTD. p. 58. ISBN 978-81-261-0675-2.

- Sumich, James L. (2008). Laboratory and Field Investigations in Marine Life. Jones & Bartlett Learning. p. 67. ISBN 978-0-7637-5730-4.

- Prewitt, Nancy L.; Larry S. Underwood; William Surver (2003). BioInquiry: making connections in biology. John Wiley. p. 289. ISBN 978-0-471-20228-8.

- Ponder, W.F.; Lindberg, D.R., eds. (2008). Phylogeny and Evolution of the Mollusca. Berkeley: University of California Press. p. 481. ISBN 978-0-520-25092-5.

- Karleskint G, Richard Turner R and , James Small J (2012) Introduction to Marine Biology Cengage Learning, edition 4, page 445. ISBN 9781133364467.

- Ruppert, Edward E.; Fox, Richard, S.; Barnes, Robert D. (2004). Invertebrate Zoology, 7th edition. Cengage Learning. ISBN 81-315-0104-3.

- Healy, J.M. (2001). "The Mollusca". In Anderson, D.T. Invertebrate Zoology (2 ed.). Oxford University Press. pp. 120–171. ISBN 0-19-551368-1.

- Raup, David M. & Stanley, Steven M. (1978). Principles of Paleontology (2 ed.). W.H. Freeman and Co. pp. 4–5. ISBN 0716700220.

- Teng, L. and Labosky P. A. (2006). "Neural crest stem cells" In: Jean-Pierre Saint-Jeannet, Neural Crest Induction and Differentiation, pp. 206-212, Springer Science & Business Media. ISBN 9780387469546.

- Richards, O. W.; Davies, R.G. (1977). Imms' General Textbook of Entomology: Volume 1: Structure, Physiology and Development Volume 2: Classification and Biology. Berlin: Springer. ISBN 0-412-61390-5.

- Langstroth, Lovell; Libby Langstroth; Todd Newberry; Monterey Bay Aquarium (2000). A living bay: the underwater world of Monterey Bay. University of California Press. p. 244. ISBN 978-0-520-22149-9.

Reproduction in Animals

Reproduction is the activity of creating newer individuals of the same species for species survival and genetic propagation. Two types of reproduction exist in animals: sexual reproduction and asexual reproduction. This chapter deals with these themes as well as related ones such as sexual dimorphism and sexual mimicry. Zoology is best understood in confluence with the major topics listed in the following chapter.

Reproduction

Reproduction (or procreation, breeding) is the biological process by which new individual organisms – "offspring" – are produced from their "parents". Reproduction is a fundamental feature of all known life; each individual organism exists as the result of reproduction. There are two forms of reproduction: asexual and sexual.

Production of new individuals along a leaf margin of the miracle leaf plant (*Kalanchoe pinnata*). The small plant in front is about 1 cm (0.4 in) tall. The concept of "individual" is obviously stretched by this asexual reproductive process.

In asexual reproduction, an organism can reproduce without the involvement of another organism. Asexual reproduction is not limited to single-celled organisms. The cloning of an organism is a form of asexual reproduction. By asexual reproduction, an organism creates a genetically similar or identical copy of itself. The evolution of sexual reproduction is a major puzzle for biologists. The two-fold cost of sex is that only 50% of organisms reproduce and organisms only pass on 50% of their genes.

Sexual reproduction typically requires the sexual interaction of two specialized organisms, called gametes, which contain half the number of chromosomes of normal cells and are created by meiosis, with typically a male fertilizing a female of the same species to create a fertilized zygote. This produces offspring organisms whose genetic characteristics are derived from those of the two parental organisms.

Asexual

Asexual reproduction is a process by which organisms create genetically similar or identical copies of themselves without the contribution of genetic material from another organism. Bacteria divide asexually via binary fission; viruses take control of host cells to produce more viruses; Hydras (invertebrates of the order *Hydroidea*) and yeasts are able to reproduce by budding. These organisms often do not possess different sexes, and they are capable of "splitting" themselves into two or more copies of themselves. Most plants have the ability to reproduce asexually and the ant species Mycocepurus smithii is thought to reproduce entirely by asexual means.

Some species that are capable of reproducing asexually, like hydra, yeast and jellyfish, may also reproduce sexually. For instance, most plants are capable of vegetative reproduction—reproduction without seeds or spores—but can also reproduce sexually. Likewise, bacteria may exchange genetic information by conjugation.

Other ways of asexual reproduction include parthenogenesis, fragmentation and spore formation that involves only mitosis. Parthenogenesis is the growth and development of embryo or seed without fertilization by a male. Parthenogenesis occurs naturally in some species, including lower plants (where it is called apomixis), invertebrates (e.g. water fleas, aphids, some bees and parasitic wasps), and vertebrates (e.g. some reptiles, fish, and, very rarely, birds and sharks). It is sometimes also used to describe reproduction modes in hermaphroditic species which can self-fertilize.

Sexual

Hoverflies mating in midair flight

Sexual reproduction is a biological process that creates a new organism by combining the genetic material of two organisms in a process that starts with meiosis, a specialized type of cell division. Each of two parent organisms contributes half of the offspring's genetic makeup by creating haploid gametes. Most organisms form two different types of gametes. In these *anisogamous* species, the two sexes are referred to as male (producing sperm or microspores) and female (producing ova or megaspores). In *isogamous species*, the gametes are similar or identical in form (isogametes), but may have separable properties and then may be given other different names. For example, in the green alga, *Chlamydomonas reinhardtii*, there are so-called "plus" and "minus" gametes. A few types of organisms, such as ciliates, *Paramecium aurelia*, have more than two types of "sex", called syngens.

Most animals (including humans) and plants reproduce sexually. Sexually reproducing organisms have different sets of genes for every trait (called alleles). Offspring inherit one allele for each trait from each parent, thereby ensuring that offspring have a combination of the parents' genes. Diploid having two copies of every gene within an organism, it is believed that "the masking of deleterious alleles favors the evolution of a dominant diploid phase in organisms that alternate between haploid and diploid phases" where recombination occurs freely.

Bryophyte reproduces sexually but its commonly seen life forms are all haploid, which produce gametes. The zygotes of the gametes develop into sporangium, which produces haploid spores. The diploid stage is relatively short compared with that of haploid stage, i.e. *haploid dominance*. The advantage of diploid, e.g. heterosis, only takes place in diploid life stage. Bryophyte still maintains the sexual reproduction during its evolution despite the fact that the haploid stage does not benefit from heterosis at all. This may be an example that the sexual reproduction has a bigger advantage by itself, since it allows gene shuffling (hybrid or recombination between multiple loci) among different members of the species, that permits natural selection of the fit over these new hybrids or recombinants that are haploid forms.

Allogamy

Allogamy is the fertilization of an ovum from one individual with the spermatozoa of another.

Autogamy

Self-fertilization, also known as autogamy, occurs in hermaphroditic organisms where the two gametes fused in fertilization come from the same individual, e.g., some foraminiferans, some ciliates. The term "autogamy" is sometimes substituted for autogamous pollination (not necessarily leading to successful fertilization) and describes self-pollination within the same flower, distinguished from geitonogamous pollination, transfer of pollen to a different flower on the same flowering plant, or within a single monoecious Gymnosperm plant. For example, species *Helonias bullata* suffer from low genetic diversity due to self-fertilization.

Mitosis and Meiosis

Mitosis and meiosis are types of cell division. Mitosis occurs in somatic cells, while meiosis occurs in gametes.

Mitosis The resultant number of cells in mitosis is twice the number of original cells. The number of chromosomes in the offspring cells is the same as that of the parent cell.

Meiosis The resultant number of cells is four times the number of original cells. This results in cells with half the number of chromosomes present in the parent cell. A diploid cell duplicates itself, then undergoes two divisions (tetraploid to diploid to haploid), in the process forming four haploid cells. This process occurs in two phases, meiosis I and meiosis II.

Same-sex

In recent decades, developmental biologists have been researching and developing techniques to facilitate same-sex reproduction. The obvious approaches, subject to a growing amount of activity, are female sperm and male eggs, with female sperm closer to being a reality for humans, given that Japanese scientists have already created female sperm for chickens. "However, the ratio of produced W chromosome-bearing (W-bearing) spermatozoa fell substantially below expectations. It is therefore concluded that most of the W-bearing PGC could not differentiate into spermatozoa because of restricted spermatogenesis." In 2004, by altering the function of a few genes involved with imprinting, other Japanese scientists combined two mouse eggs to produce daughter mice.

Strategies

There are a wide range of reproductive strategies employed by different species. Some animals, such as the human and northern gannet, do not reach sexual maturity for many years after birth and even then produce few offspring. Others reproduce quickly; but, under normal circumstances, most offspring do not survive to adulthood. For example, a rabbit (mature after 8 months) can produce 10–30 offspring per year, and a fruit fly (mature after 10–14 days) can produce up to 900 offspring per year. These two main strategies are known as K-selection (few offspring) and r-selection (many offspring). Which strategy is favoured by evolution depends on a variety of circumstances. Animals with few offspring can devote more resources to the nurturing and protection of each individual offspring, thus reducing the need for many offspring. On the other hand, animals with many offspring may devote fewer resources to each individual offspring; for these types of animals it is common for many offspring to die soon after birth, but enough individuals typically survive to maintain the population. Some organisms such as honey bees and fruit flies retain sperm in a process called sperm storage thereby increasing the duration of their fertility.

Other Types

- Polycyclic animals reproduce intermittently throughout their lives.

- Semelparous organisms reproduce only once in their lifetime, such as annual plants (including all grain crops), and certain species of salmon, spider, bamboo and century plant. Often, they die shortly after reproduction. This is often associated with r-strategists.

- Iteroparous organisms produce offspring in successive (e.g. annual or seasonal) cycles, such as perennial plants. Iteroparous animals survive over multiple seasons (or periodic condition changes). This is more associated with K-strategists.

Asexual vs. Sexual Reproduction

Organisms that reproduce through asexual reproduction tend to grow in number exponentially. However, because they rely on mutation for variations in their DNA, all members of the species have similar vulnerabilities. Organisms that reproduce sexually yield a smaller number of offspring, but the large amount of variation in their genes makes them less susceptible to disease.

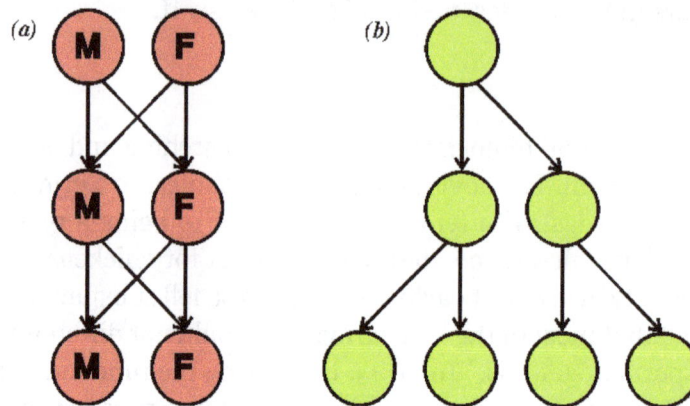

Illustration of the *twofold cost of sexual reproduction*. If each organism were to contribute to the same number of offspring (two), *(a)* the population remains the same size each generation, where the *(b)* asexual population doubles in size each generation.

Many organisms can reproduce sexually as well as asexually. Aphids, slime molds, sea anemones, some species of starfish (by fragmentation), and many plants are examples. When environmental factors are favorable, asexual reproduction is employed to exploit suitable conditions for survival such as an abundant food supply, adequate shelter, favorable climate, disease, optimum pH or a proper mix of other lifestyle requirements. Populations of these organisms increase exponentially via asexual reproductive strategies to take full advantage of the rich supply resources.

When food sources have been depleted, the climate becomes hostile, or individual survival is jeopardized by some other adverse change in living conditions, these organisms switch to sexual forms of reproduction. Sexual reproduction ensures a mixing of the gene pool of the species. The variations found in offspring of sexual reproduction allow some individuals to be better suited for survival and provide a mechanism for selective adaptation to occur. The meiosis stage of the sexual cycle also allows especially effective repair of DNA damages. In addition, sexual reproduction usually results in the formation of a life stage that is able to endure the conditions that threaten the offspring of an asexual parent. Thus, seeds, spores, eggs, pupae, cysts or other "over-wintering" stages of sexual reproduction ensure the survival during unfavorable times and the organism can "wait out" adverse situations until a swing back to suitability occurs.

Life Without

The existence of life without reproduction is the subject of some speculation. The biological study of how the origin of life produced reproducing organisms from non-reproducing elements is called abiogenesis. Whether or not there were several independent abiogenetic events, biologists believe that the last universal ancestor to all present life on Earth lived about 3.5 billion years ago.

Scientists have speculated about the possibility of creating life non-reproductively in the laboratory. Several scientists have succeeded in producing simple viruses from entirely non-living materials. However, viruses are often regarded as not alive. Being nothing more than a bit of RNA or DNA in a protein capsule, they have no metabolism and can only replicate with the assistance of a hijacked cell's metabolic machinery.

The production of a truly living organism (e.g. a simple bacterium) with no ancestors would be a much more complex task, but may well be possible to some degree according to current biological knowledge. A synthetic genome has been transferred into an existing bacterium where it replaced the native DNA, resulting in the artificial production of a new *M. mycoides* organism.

There is some debate within the scientific community over whether this cell can be considered completely synthetic on the grounds that the chemically synthesized genome was an almost 1:1 copy of a naturally occurring genome and, the recipient cell was a naturally occurring bacterium. The Craig Venter Institute maintains the term "synthetic bacterial cell" but they also clarify "...we do not consider this to be "creating life from scratch" but rather we are creating new life out of already existing life using synthetic DNA". Venter plans to patent his experimental cells, stating that "they are pretty clearly human inventions". Its creators suggests that building 'synthetic life' would allow researchers to learn about life by building it, rather than by tearing it apart. They also propose to stretch the boundaries between life and machines until the two overlap to yield "truly programmable organisms". Researchers involved stated that the creation of "true synthetic biochemical life" is relatively close in reach with current technology and cheap compared to the effort needed to place man on the Moon.

Lottery Principle

Sexual reproduction has many drawbacks, since it requires far more energy than asexual reproduction and diverts the organisms from other pursuits, and there is some argument about why so many species use it. George C. Williams used lottery tickets as an analogy in one explanation for the widespread use of sexual reproduction. He argued that asexual reproduction, which produces little or no genetic variety in offspring, was like buying many tickets that all have the same number, limiting the chance of "winning" - that is, producing surviving offspring. Sexual reproduction, he argued, was like purchasing fewer tickets but with a greater variety of numbers and therefore a greater chance of success. The point of this analogy is that since asexual reproduction does not produce genetic variations, there is little ability to quickly adapt to a changing environment. The lottery principle is less accepted these days because of evidence that asexual reproduction is more prevalent in unstable environments, the opposite of what it predicts.

Sexual Reproduction

Sexual reproduction is a form of reproduction where two morphologically distinct types of specialized reproductive cells called gametes fuse together, involving a female's large ovum (or egg) and a male's smaller sperm. Each gamete contains half the number of chromosomes of normal cells. They are created by a specialized type of cell division, which only occurs in eukaryotic cells, known as meiosis. The two gametes fuse during fertilization to produce DNA replication and the creation of a single-celled zygote which includes genetic material from both gametes. In a process called genetic recombination, genetic material (DNA) joins up so that homologous chromosome sequences are aligned with each other, and this is followed by exchange of genetic information. Two rounds of cell division then produce four daughter cells with half the number of chromosomes from each original parent cell, and the same number of chromosomes as both parents, though self-fertilization can occur. For instance, in human reproduction each human cell contains 46 chromosomes,

23 pairs, except gamete cells, which only contain 23 chromosomes, so the child will have 23 chromosomes from each parent genetically recombined into 23 pairs. Cell division initiates the development of a new individual organism in multicellular organisms, including animals and plants, for the vast majority of whom this is the primary method of reproduction. A species is defined as a taxonomic rank. A species is often defined as the largest group of organisms where two hybrids are capable of reproducing fertile offspring, typically using sexual reproduction, although the species problem encompasses a series of difficult related questions that often come up when biologists define the word species.

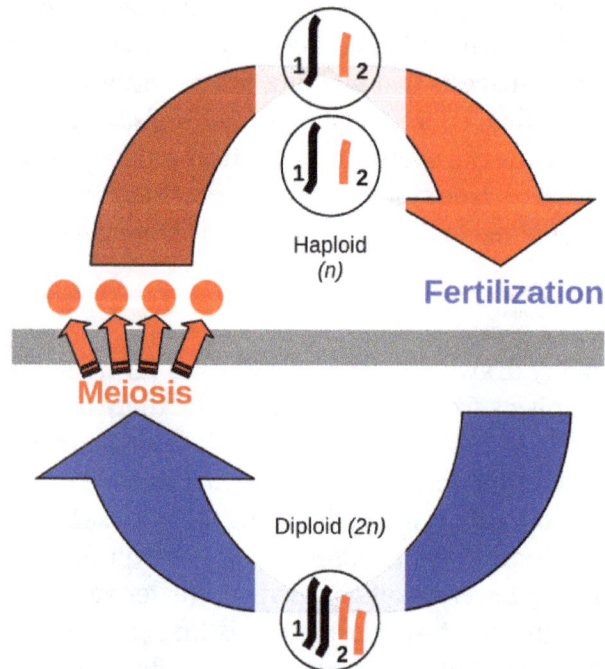

In the first stage of sexual reproduction, "meiosis", the number of chromosomes is reduced from a diploid number (2n) to a haploid number (n). During "fertilization", haploid gametes come together to form a diploid zygote and the original number of chromosomes is restored.

The evolution of sexual reproduction is a major puzzle because asexual reproduction should be able to outcompete it as every young organism created can bear its own young. This implies that an asexual population has an intrinsic capacity to grow more rapidly with each generation. This 50% cost is a fitness disadvantage of sexual reproduction. The two-fold cost of sex includes this cost and the fact that any organism can only pass on 50% of its own genes to its offspring. One definite advantage of sexual reproduction is that it prevents the accumulation of genetic mutations.

Sexual selection is a mode of natural selection in which some individuals out-reproduce others of a population because they are better at securing mates for sexual reproduction. It has been described as "a powerful evolutionary force that does not exist in asexual populations."

Prokaryotes, whose initial cell has additional or transformed genetic material, reproduce through asexual reproduction but may, in lateral gene transfer, display processes such as bacterial conjugation, transformation and transduction, which are similar to sexual reproduction although they do not lead to reproduction.

Evolution

The first fossilized evidence of sexual reproduction in eukaryotes is from the Stenian period, about 1 to 1.2 billion years ago.

Biologists studying evolution propose several explanations for why sexual reproduction developed and why it is maintained. These reasons include fighting the accumulation of deleterious mutations, increasing rate of adaptation to changing environments, dealing with competition, or masking deleterious mutations. All of these ideas about why sexual reproduction has been maintained are generally supported, but ultimately the size of the population determines if sexual reproduction is entirely beneficial. Larger populations appear to respond more quickly to benefits obtained through sexual reproduction than do smaller population sizes.

Maintenance of sexual reproduction has been explained by theories that work at several levels of selection, though some of these models remain controversial.

New models presented in recent years suggest a basic advantage for sexual reproduction in slowly reproducing complex organisms. Sexual reproduction allows these species to exhibit characteristics that depend on the specific environment that they inhabit, and the particular survival strategies that they employ.

Sexual Selection

In order to sexually reproduce, both males and females need to find a mate. Generally in animals mate choice is made by females while males compete to be chosen. This can lead organisms to extreme efforts in order to reproduce, such as combat and display, or produce extreme features caused by a positive feedback known as a Fisherian runaway. Thus sexual reproduction, as a form of natural selection, has an effect on evolution. Sexual dimorphism is where the basic phenotypic traits vary between males and females of the same species. Dimorphism is found in both sex organs and in secondary sex characteristics, body size, physical strength and morphology, biological ornamentation, behavior and other bodily traits. However, sexual selection is only implied over an extended period of time leading to sexual dimorphism.

Sex Ratio

Apart from some eusocial wasps, organisms which reproduce sexually have a 1:1 sex ratio of male and female births. The English statistician and biologist Ronald Fisher outlined why this is so in what has come to be known as Fisher's principle. This essentially says the following:

1. Suppose male births are less common than female.

2. A newborn male then has better mating prospects than a newborn female, and therefore can expect to have more offspring.

3. Therefore parents genetically disposed to produce males tend to have more than average numbers of grandchildren born to them.

4. Therefore the genes for male-producing tendencies spread, and male births become more common.

5. As the 1:1 sex ratio is approached, the advantage associated with producing males dies away.

6. The same reasoning holds if females are substituted for males throughout. Therefore 1:1 is the equilibrium ratio.

Animals

Insects

Australian emperor laying egg, guarded by the male

Insect species make up more than two-thirds of all extant animal species. Most insect species reproduce sexually, though some species are facultatively parthenogenetic. Many insects species have sexual dimorphism, while in others the sexes look nearly identical. Typically they have two sexes with males producing spermatozoa and females ova. The ova develop into eggs that have a covering called the chorion, which forms before internal fertilization. Insects have very diverse mating and reproductive strategies most often resulting in the male depositing spermatophore within the female, which she stores until she is ready for egg fertilization. After fertilization, and the formation of a zygote, and varying degrees of development, in many species the eggs are deposited outside the female; while in others, they develop further within the female and are born live.

Birds

Mammals

There are three extant kinds of mammals: monotremes, placentals and marsupials, all with internal fertilization. In placental mammals, offspring are born as juveniles: complete animals with the sex organs present although not reproductively functional. After several months or years, depending on the species, the sex organs develop further to maturity and the animal becomes sexually mature. Most female mammals are only fertile during certain periods during their estrous cycle, at which point they are ready to mate. Individual male and female mammals meet and carry out copulation. For most mammals, males and females exchange sexual partners throughout their adult lives.

Fish

The vast majority of fish species lay eggs that are then fertilized by the male, some species lay their eggs on a substrate like a rock or on plants, while others scatter their eggs and the eggs are fertilized as they drift or sink in the water column.

Some fish species use internal fertilization and then disperse the developing eggs or give birth to live offspring. Fish that have live-bearing offspring include the guppy and mollies or Poecilia. Fishes that give birth to live young can be ovoviviparous, where the eggs are fertilized within the female and the eggs simply hatch within the female body, or in seahorses, the male carries the developing young within a pouch, and gives birth to live young. Fishes can also be viviparous, where the female supplies nourishment to the internally growing offspring. Some fish are hermaphrodites, where a single fish is both male and female and can produce eggs and sperm. In hermaphroditic fish, some are male and female at the same time while in other fish they are serially hermaphroditic; starting as one sex and changing to the other. In at least one hermaphroditic species, self-fertilization occurs when the eggs and sperm are released together. Internal self-fertilization may occur in some other species. One fish species does not reproduce by sexual reproduction but uses sex to produce offspring; Poecilia formosa is a unisex species that uses a form of parthenogenesis called gynogenesis, where unfertilized eggs develop into embryos that produce female offspring. Poecilia formosa mate with males of other fish species that use internal fertilization, the sperm does not fertilize the eggs but stimulates the growth of the eggs which develops into embryos.

Reptiles

Plants

Animals typically produce gametes directly by meiosis. Male gametes are called sperm, and female gametes are called eggs or ova. In animals, fertilization follows immediately after meiosis. Plants on the other hand have mitosis occurring in spores, which are produced by meiosis. The spores germinate into the gametophyte phase. The gametophytes of different groups of plants vary in size; angiosperms have as few as three cells in pollen, and mosses and other so called primitive plants may have several million cells. Plants have an alternation of generations where the sporophyte phase is succeeded by the gametophyte phase. The sporophyte phase produces spores within the sporangium by meiosis.

Flowering Plants

Flowering plants are the dominant plant form on land and they reproduce either sexually or asexually. Often their most distinguishing feature is their reproductive organs, commonly called flowers. The anther produces pollen grains which contain the male gametophytes (sperm). For pollination to occur, pollen grains must attach to the stigma of the female reproductive structure (carpel), where the female gametophytes (ovules) are located inside the ovary. After the pollen tube grows through the carpel's style, the sex cell nuclei from the pollen grain migrate into the ovule to fertilize the egg cell and endosperm nuclei within the female gametophyte in a process termed double fertilization. The resulting zygote develops into an embryo, while the triploid endosperm (one sperm cell plus two female cells) and female tissues of the ovule give rise to the surrounding tissues in the developing seed. The ovary, which produced the female gametophyte(s), then grows into a fruit, which surrounds the seed(s). Plants may either self-pollinate or cross-pollinate.

Flowers are the sexual organs of flowering plants.

Nonflowering plants like ferns, moss and liverworts use other means of sexual reproduction.

In 2013, flowers dating from the Cretaceous (100 million years before present) were found encased in amber, the oldest evidence of sexual reproduction in a flowering plant. Microscopic images showed tubes growing out of pollen and penetrating the flower's stigma. The pollen was sticky, suggesting it was carried by insects.

Ferns

Ferns mostly produce large diploid sporophytes with rhizomes, roots and leaves; and on fertile leaves called sporangium, spores are produced. The spores are released and germinate to produce short, thin gametophytes that are typically heart shaped, small and green in color. The gametophytes or thallus, produce both motile sperm in the antheridia and egg cells in separate archegonia. After rains or when dew deposits a film of water, the motile sperm are splashed away from the antheridia, which are normally produced on the top side of the thallus, and swim in the film of water to the archegonia where they fertilize the egg. To promote out crossing or cross fertilization the sperm are released before the eggs are receptive of the sperm, making it more likely that the sperm will fertilize the eggs of different thallus. A zygote is formed after fertilization, which grows into a new sporophytic plant. The condition of having separate sporephyte and gametophyte plants is called alternation of generations. Other plants with similar reproductive means include the *Psilotum*, *Lycopodium*, *Selaginella* and *Equisetum*.

Bryophytes

The bryophytes, which include liverworts, hornworts and mosses, reproduce both sexually and vegetatively. They are small plants found growing in moist locations and like ferns, have motile sperm with flagella and need water to facilitate sexual reproduction. These plants start as a haploid spore that grows into the dominate form, which is a multicellular haploid body with leaf-like structures that photosynthesize. Haploid gametes are produced in antherida and archegonia by mitosis. The sperm released from the antherida respond to chemicals released by ripe archegonia and swim to them in a film of water and fertilize the egg cells thus producing a zygote. The zygote divides by mitotic division and grows into a sporophyte that is diploid. The multicellular diploid sporophyte produces structures called spore capsules, which are connected by seta to the archegonia. The spore capsules produce spores by meiosis, when ripe the capsules burst open and the spores are released. Bryophytes show considerable variation in their breeding structures and the above is a basic outline. Also in some species each plant is one sex while other species produce both sexes on the same plant.

Fungi

Fungi are classified by the methods of sexual reproduction they employ. The outcome of sexual reproduction most often is the production of resting spores that are used to survive inclement times and to spread. There are typically three phases in the sexual reproduction of fungi: plasmogamy, karyogamy and meiosis.

Bacteria and Archaea

Three distinct processes in prokaryotes are regarded as similar to eukaryotic sex: bacterial transformation, which involves the incorporation of foreign DNA into the bacterial chromosome; bacterial conjugation, which is a transfer of plasmid DNA between bacteria, but the plasmids are rarely incorporated into the bacterial chromosome; and gene transfer and genetic exchange in archaea.

Bacterial transformation involves the recombination of genetic material and its function is mainly associated with DNA repair. Bacterial transformation is a complex process encoded by numerous bacterial genes, and is a bacterial adaptation for DNA transfer. This process occurs naturally in at least 40 bacterial species. For a bacterium to bind, take up, and recombine exogenous DNA into its chromosome, it must enter a special physiological state referred to as competence. Sexual reproduction in early single-celled eukaryotes may have evolved from bacterial transformation, or from a similar process in archaea.

On the other hand, bacterial conjugation is a type of direct transfer of DNA between two bacteria through an external appendage called the conjugation pilus. Bacterial conjugation is controlled by plasmid genes that are adapted for spreading copies of the plasmid between bacteria. The infrequent integration of a plasmid into a host bacterial chromosome, and the subsequent transfer of a part of the host chromosome to another cell do not appear to be bacterial adaptations.

Exposure of hyperthermophilic archaeal Sulfolobus species to DNA damaging conditions induces cellular aggregation accompanied by high frequency genetic marker exchange. Ajon et al. hypothesized that this cellular aggregation enhances species-specific DNA repair by homologous recombination. DNA transfer in Sulfolobus may be an early form of sexual interaction similar to the more well-studied bacterial transformation systems that also involve species-specific DNA transfer leading to homologous recombinational repair of DNA damage.

Asexual Reproduction

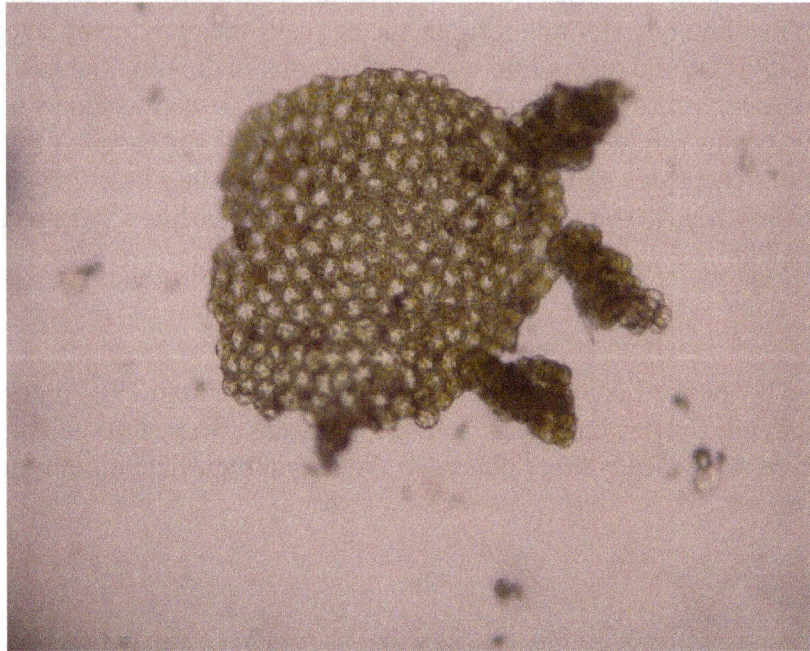

Asexual reproduction in liverworts: a caducous phylloid germinating

Asexual reproduction is a type of reproduction by which offspring arise from a single organism, and inherit the genes of that parent only; it does not involve the fusion of gametes and almost never changes the number of chromosomes. Asexual reproduction is the primary form of reproduction for single-celled organism as the archaea, bacteria, and protists. Many plants and fungi reproduce asexually as well.

While all prokaryotes reproduce asexually (without the formation and fusion of gametes), mechanisms for lateral gene transfer such as conjugation, transformation and transduction are sometimes likened to sexual reproduction (or at least with sex, in the sense of genetic recombination). A complete lack of sexual reproduction is relatively rare among multicellular organisms, particularly animals. It is not entirely understood why the ability to reproduce sexually is so common among them. Current hypotheses suggest that asexual reproduction may have short term benefits when rapid population growth is important or in stable environments, while sexual reproduction offers a net advantage by allowing more rapid generation of genetic diversity, allowing adaptation to changing environments. Developmental constraints may underlie why few animals have relinquished sexual reproduction completely in their life-cycles. Another constraint on switching from sexual to asexual reproduction would be the concomitant loss of meiosis and the protective recombinational repair of DNA damage afforded as one function of meiosis.

Types

Binary Fission

An important form of fission is binary fission. In binary fission, the parent organism is replaced by two daughter organisms, because it literally divides in two. Only prokaryotes (the archaea and the bacteria) reproduce asexually through binary fission. Eukaryotes (such as protists and unicellular fungi) reproduce by mitosis; most of these are also capable of sexual reproduction.

Another type of fission is multiple fission that is advantageous to the plant life cycle. Multiple fission at the cellular level occurs in many protists, e.g. sporozoans and algae. The nucleus of the parent cell divides several times by mitosis, producing several nuclei. The cytoplasm then separates, creating multiple daughter cells.

In apicomplexans, multiple fission, or schizogony, is manifested either as merogony, sporogony or gametogony. Merogony results in merozoites, which are multiple daughter cells, that originate within the same cell membrane, sporogony results in sporozoites, and gametogony results in microgametes.

Budding

Some cells split via budding (for example baker's yeast), resulting in a "mother" and "daughter" cell. The offspring organism is smaller than the parent. Budding is also known on a multicellular level; an animal example is the hydra, which reproduces by budding. The buds grow into fully matured individuals which eventually break away from the parent organism.

Internal budding or Endodyogeny is a process of asexual reproduction, favoured by parasites such as *Toxoplasma gondii*. It involves an unusual process in which two daughter cells are produced inside a mother cell, which is then consumed by the offspring prior to their separation.

Endopolygeny is the division into several organisms at once by internal budding. Also, budding (external or internal) is present in some worm like Taenia or Echinococci; these worm produce cyst and then produce (invaginated or evaginated) protoscolex with budding.

Vegetative Reproduction

Closeup of a Bryophyllum daigremontianum

Vegetative reproduction is a type of asexual reproduction found in plants where new individuals are formed without the production of seeds or spores by meiosis or syngamy. Examples of vegetative reproduction include the formation of miniaturized plants called plantlets on specialized leaves (for example in kalanchoe) and some produce new plants out of rhizomes or stolon (for example in strawberry). Other plants reproduce by forming bulbs or tubers (for example tulip bulbs and dahlia tubers). Some plants produce adventitious shoots and omay form a clonal colony, where all the individuals are clones, and the clones may cover a large area.

Spore

Many multicellular organisms form spores during their biological life cycle in a process called *sporogenesis*. Exceptions are animals and some protists, who undergo *meiosis* immediately followed by fertilization. Plants and many algae on the other hand undergo *sporic meiosis* where meiosis leads to the formation of haploid spores rather than gametes. These spores grow into multicellular individuals (called gametophytes in the case of plants) without a fertilization event. These haploid individuals give rise to gametes through mitosis. Meiosis and gamete formation therefore occur in separate generations or "phases" of the life cycle, referred to as alternation of generations. Since sexual reproduction is often more narrowly defined as the fusion of gametes (fertilization), spore formation in plant sporophytes and algae might be considered a form of asexual reproduction (agamogenesis) despite being the result of meiosis and undergoing a reduction in ploidy. However, both events (spore formation and fertilization) are necessary to complete sexual reproduction in the plant life cycle.

Fungi and some algae can also utilize true asexual spore formation, which involves mitosis giving rise to reproductive cells called mitospores that develop into a new organism after dispersal. This method of reproduction is found for example in conidial fungi and the red algae *Polysiphonia*, and involves sporogenesis without meiosis. Thus the chromosome number of the spore cell is the same as that of the parent producing the spores. However, mitotic sporogenesis is an exception and most spores, such as those of plants, most Basidiomycota, and many algae, are produced by meiosis.

Fragmentation

Fragmentation is a form of asexual reproduction where a new organism grows from a fragment of the parent. Each fragment develops into a mature, fully grown individual. Fragmentation is seen in many organisms such as animals (some planarian and annelid worms,[which?] turbellarians and sea stars), fungi, and plants. Some plants have specialized structures for reproduction via fragmentation, such as *gemma* in liverworts. Most lichens, which are a symbiotic union of a fungus and photosynthetic algae or bacteria, reproduce through fragmentation to ensure that new individuals contain both symbiont. These fragments can take the form of *soredia*, dust-like particles consisting of fungal hyphen wrapped around photobiont cells.

Clonal Fragmentation in multicellular or colonial organisms is a form of asexual reproduction or cloning where an organism is split into fragments. Each of these fragments develop into mature, fully grown individuals that are clones of the original organism. In echinoderms, this method of reproduction is usually known as fissiparity. Researchers claim today that due to many environmental and epigenetic differences, that clones originated in the same ancestor might actually be genetically and epigenetically different.

Agamogenesis

Agamogenesis is any form of reproduction that does not involve a male gamete. Examples are parthenogenesis and apomixis.

Parthenogenesis

Parthenogenesis is a form of agamogenesis in which an unfertilized egg develops into a new individual. Parthenogenesis occurs naturally in many plants, invertebrates (e.g. water fleas, rotifers, aphids, stick insects, some ants, bees and parasitic wasps), and vertebrates (e.g. some reptiles, amphibians, rarely birds). In plants, apomixis may or may not involve parthenogenesis.

Apomixis and Nucellar Embryony

Apomixis in plants is the formation of a new sporophyte without fertilization. It is important in ferns and in flowering plants, but is very rare in other seed plants. In flowering plants, the term "apomixis" is now most often used for agamospermy, the formation of seeds without fertilization, but was once used to include vegetative reproduction. An example of an apomictic plant would be the triploid European dandelion. Apomixis mainly occurs in two forms: In gametophytic apomixis, the embryo arises from an unfertilized egg within a diploid embryo sac that was formed without completing meiosis. In nucellar embryony, the embryo is formed from the diploid nucellus tissue surrounding the embryo sac. Nucellar embryony occurs in some citrus seeds. Male apomixis can occur in rare cases, such as the Saharan Cypress *Cupressus dupreziana*, where the genetic material of the embryo are derived entirely from pollen. The term "apomixis" is also used for asexual reproduction in some animals, notably water-fleas, *Daphnia*.

Alternation between Sexual and Asexual Reproduction

Some species alternate between the sexual and asexual strategies, an ability known as heterogamy, depending on conditions. Alternation is observed in several rotifer species (cyclical parthenogenesis e.g. in Brachionus species) and a few types of insects, such as aphids which will, under certain conditions, produce eggs that have not gone through meiosis, thus cloning themselves. The cape bee *Apis mellifera* subsp. *capensis* can reproduce asexually through a process called thelytoky. A few species of amphibians, reptiles, and birds have a similar ability. For example, the freshwater crustacean *Daphnia* reproduces by parthenogenesis in the spring to rapidly populate ponds, then switches to sexual reproduction as the intensity of competition and predation increases. Another example are monogonont rotifers of the genus *Brachionus*, which reproduce via cyclical parthenogenesis: at low population densities females produce asexually and at higher densities a chemical cue accumulates and induces the transition to sexual reproduction. Many protists and fungi alternate between sexual and asexual reproduction.

For example, the slime mold *Dictyostelium* undergoes binary fission (mitosis) as single-celled amoebae under favorable conditions. However, when conditions turn unfavorable, the cells aggregate and follow one of two different developmental pathways, depending on conditions. In the social pathway, they form a multicellular slug which then forms a fruiting body with asexually generated spores. In the sexual pathway, two cells fuse to form a giant cell that develops into a large cyst. When this macrocyst germinates, it releases hundreds of amoebic cells that are the product of meiotic recombination between the original two cells.

The hyphae of the common mold (*Rhizopus*) are capable of producing both mitotic as well as meiotic spores. Many algae similarly switch between sexual and asexual reproduction. A number of plants use both sexual and asexual means to produce new plants, some species alter their primary modes of reproduction from sexual to asexual under varying environmental conditions.

Inheritance in Sexual Species

For example, in the rotifer *Brachionus calyciflorus* asexual reproduction (obligate parthenogenesis) can be inherited by a recessive allele, which leads to loss of sexual reproduction in homozygous offspring. Inheritance of asexual reproduction by a single recessive locus has also been found in the parasitoid wasp *Lysiphlebus fabarum*.

Examples in Animals

There are examples of parthenogenesis in the hammerhead shark and the blacktip shark. In both cases, the sharks had reached sexual maturity in captivity in the absence of males, and in both cases the offspring were shown to be genetically identical to the mothers. The New Mexico whiptail is another example.

Reptiles use the ZW sex-determination system, which produces either males (with ZZ sex chromosomes) or females (with ZW or WW sex chromosomes). Until 2010, it was thought that the ZW chromosome system used by reptiles was incapable of producing viable WW offspring, but a (ZW) female boa constrictor was discovered to have produced viable female offspring with WW chromosomes. The female boa could have chosen any number of male partners (and had successfully in the past) but on these occasions she reproduced asexually, creating 22 female babies with WW sex-chromosomes.

Polyembryony is a widespread form of asexual reproduction in animals, whereby the fertilized egg or a later stage of embryonic development splits to form genetically identical clones. Within animals, this phenomenon has been best studied in the parasitic Hymenoptera. In the 9-banded armadillos, this process is obligatory and usually gives rise to genetically identical quadruplets. In other mammals, monozygotic twinning has no apparent genetic basis, though its occurrence is common. There are at least 10 million identical human twins and triplets in the world today.

Bdelloid rotifers reproduce exclusively asexually, and all individuals in the class Bdelloidea are females. Asexuality evolved in these animals millions of years ago and has persisted since. There is evidence to suggest that asexual reproduction has allowed the animals to evolve new proteins through the Meselson effect that have allowed them to survive better in periods of dehydration.

Molecular evidence strongly suggest that several species of the stick insect genus *Timema* have used only asexual (parthenogenetic) reproduction for millions of years, the longest period known for any insect.

In the grass thrips genus *Aptinothrips* there have been several transitions to asexuality, likely due to different causes.

Sexual Dimorphism

Female (left) and male (right) common pheasant, illustrating the dramatic difference in both color and size between sexes

Sexual dimorphism is the condition where the two sexes of the same species exhibit different characteristics beyond the differences in their sexual organs. The condition occurs in many animals, insects, birds and some plants. Differences may include secondary sex characteristics, size, color, markings, and may also include behavioral differences. These differences may be subtle or exaggerated, and may be subjected to sexual selection. The opposite of dimorphism is *monomorphism*.

Overview

The peacock, on the right, is courting the peahen, on the left.

Male (bottom) and female mallards. The male mallard has an unmistakable green head.

Orgyia antiqua male and female.

Ornamentation and Coloration

Common and easily identified types of dimorphism are ornamentation and coloration, though not always apparent. A difference in coloration of sexes within a given species is called sexual dichromatism, which is commonly seen in many species of birds and reptiles. Sexual selection leads to the exaggerated dimorphic traits that are used predominantly in competition over mates. The increased fitness resulting from ornamentation offsets its cost to produce or maintain suggesting complex evolutionary implications, but the costs and evolutionary implications vary from species to species.

Exaggerated ornamental traits are used predominantly in the competition over mates, implying sexual selection. Ornaments may be costly to produce or maintain, which has complex evolutionary implications but the costs and implications differ depending on the nature of the ornamentation (such as the colour mechanism involved).

The peafowl constitute conspicuous illustrations of the principle. The ornate plumage of peacocks, as used in the courting display, attracts peahens. At first sight one might mistake peacocks and peahens for completely different species because of the vibrant colours and the sheer size of the male's plumage; the peahen being of a subdued brown coloration. The plumage of the peacock increases its vulnerability to predators because it is a hindrance in flight, and it renders the bird conspicuous in general. Similar examples are manifold, such as in birds of paradise and argus pheasants.

Another example of sexual dichromatism is that of the nestling blue tits. Males are chromatically more yellow than females. It is believed that this is obtained by the ingestion of green lepidopteran larvae, which contain large amounts of the carotenoids lutein and zeaxanthin. This diet also affects the sexually dimorphic colours in the human-invisible UV spectrum. Hence, the male birds, although appearing yellow to humans actually have a violet-tinted plumage that is seen by females. This plumage is thought to be an indicator of male parental abilities. Perhaps this is a good indicator for females because it shows that they are good at obtaining a food supply from which the carotenoid is obtained. There is a positive correlation between the chromas of the tail and breast feathers and body condition. Carotenoids play an important role in immune function for many animals, so carotenoid dependent signals might indicate health.

Frogs constitute another conspicuous illustration of the principle. There are two types of dichromatism for frog species: ontogenetic and dynamic. Ontogenetic frogs are more common and have permanent color changes in males or females. Litoria lesueuri is an example of a dynamic frog that has temporarily color changes in males during breeding season. Hyperolius ocellatus is an onto-genetic frog with dramatic differences in both color and pattern between the sexes. At sexual maturity, the males display a bright green with white dorsolateral lines. In contrast, the females are rusty red to silver with small spots. The bright coloration in the male population serves to attract females and as an aposematic sign to potential predators.

Females often show a preference for exaggerated male secondary sexual characteristics in mate selection. The sexy son hypothesis explains that females prefer more elaborate males and select against males that are dull in color, independent of the species' vision.

Similar sexual dimorphism and mating choice are also observed in many fish species. For example, in guppies males have colorful spots and ornamentations while females are generally grey in color. Female guppies prefer brightly colored males to duller males.

Physiological Differentiation

In redlip blennies, only the male fish develops an organ at the anal-urogenital region that produces antimicrobial substances. During parental care, males rub their anal-urogenital regions over their nests' internal surfaces, thereby protecting their eggs from microbial infections, one of the most common causes for mortality in young fish.

Plants

Most plants are hermaphroditic but approximately 6% have separate males and females (dioecy). Males and females in insect-pollinated species generally look similar to one another because plants provide rewards (e.g. nectar) that encourage pollinators to visit another similar flower, completing pollination. *Catasetum* orchids are one interesting exception to this rule. Male *Catasetum* orchids violently attach pollinia to euglossine bee pollinators. The bees will then avoid other male flowers but may visit the female, which looks different from the males.

Various other dioecious exceptions, such as Loxostylis alata have visibly different genders, with the effect of eliciting the most efficient behaviour from pollinators, who then use the most efficient strategy in visiting each gender of flower instead of searching say, for pollen in a nectar-bearing female flower.

Some plants, such as some species of *Geranium* have what amounts to serial sexual dimorphism. The flowers of such species might for example present their anthers on opening, then shed the exhausted anthers after a day or two and perhaps change their colours as well while the pistil matures; specialist pollinators are very much inclined to concentrate on the exact appearance of the flowers they serve, which saves their time and effort and serves the interests of the plant accordingly. Some such plants go even further and change their appearance again once they have been fertilised, thereby discouraging further visits from pollinators. This is advantageous to both parties because it avoids damage to the developing fruit and avoids wasting the pollinator's effort on unrewarding visits. In effect the strategy ensures that the pollinators can expect a reward every time they visit an appropriately advertising flower.

Females of the aquatic plant *Vallisneria americana* have floating flowers attached by a long flower stalk that are fertilized if they contact one of the thousands of free floating flowers released by a male. Sexual dimorphism is most often associated with wind-pollination in plants due to selection for efficient pollen dispersal in males vs pollen capture in females, e.g. *Leucadendron rubrum*.

Sexual dimorphism in plants can also be dependent on reproductive development. This can be seen in *Cannabis sativa*, a type of hemp, which have higher photosynthesis rates in males while growing but higher rates in females once the plants become sexually mature.

It also should be borne in mind that every sexually reproducing extant species of vascular plant actually has an alternation of generations; the plants we see about us generally are diploid sporophytes, but their offspring really are not the seeds that people commonly recognise as the new generation. The seed actually is the offspring of the haploid generation of microgametophytes (pollen) and megagametophytes (the embryo sacs in the ovules). Each pollen grain accordingly may be seen as a male plant in its own right; it produces a sperm cell and is dramatically different from the female plant, the megagametophyte that produces the female gamete.

Insects

Sexual dimorphism: *Anthocharis cardamines* male is brightly coloured.

Sexual dimorphism: *Anthocharis cardamines* female

Insects display a wide variety of sexual dimorphism between taxa including size, ornamentation and coloration. The female-biased sexual size dimorphism observed in many taxa evolved despite intense male-male competition for mates. In Osmia rufa, for example, the female is larger/broader than males, with males being 8–10 mm in size and females being 10–12 mm in size. The reason for the sexual dimorphism is due to provision size mass, in which females consume more pollen than males. In some species, there evidence of male dimorphism, but it appears to be for the purpose of distinctions of roles. This is seen in the bee species Perdita portalis in which there is a small headed morph, capable of flight, and large headed morph, incapable of flight, for males. *Anthidium manicatum* also displays male-biased sexual dimorphism. The selection for larger size in males rather than females in this species may have resulted due to their aggressive territorial behavior and subsequent differential mating success. Another example is the *L. hemichalceum*, which is a species of sweat bee that shows drastic physical dimorphisms between male offpsring. Not all dimorphism has to have a drastic difference between the sexes. *Andrena agilissima* is a mining bee where the females only have a slightly larger head than the males.

Weaponry leads to increased fitness by increasing success in male-male competition in many insect species. The beetle horns in Onthophagus taurus are enlarged growths of the head or thorax expressed only in the males. Copris ochus also has distinct sexual and male dimorphism in head horns. These structures are impressive because of the exaggerated sizes. There is a direct correlation between male horn lengths and body size and higher access to mates and fitness. In other beetle species, both males and females may have ornamentation such as horns. Generally, insect sexual size dimorphism (SSD) within species increases with body size.

Sexual dimorphism within insects is also displayed by dichromatism. In butterfly genera Bicyclus and Junonia, dimorphic wing patterns evolved due to sex-limited expression, which mediates the intralocus sexual conflict and leads to increased fitness in males. The sexual dichromatic nature of Bicyclus anyana is reflected by female selection on the basis of dorsal UV-reflective eyespot pupils (Robertson & Monteiro, 2005). Naturally selected deviation in protective female coloration is displayed in mimetic butterflies.

Spiders and Sexual Cannibalism

Female (left) and Male (right) *Argiope appensa*, displaying typical sexual differences in spiders, with dramatically smaller males

Size dimorphism shows a correlation with sexual cannibalism, which is prominent in spiders (it is also found in insects such as praying mantises). In the size dimorphic wolf spider, food-limited females cannibalize more frequently. Therefore, there is a high risk of low fitness for males due to pre-copulatory cannibalism, which led to male selection of larger females for two reasons: higher fecundity and lower rates of cannibalism. In addition, female fecundity is positively correlated with female body size and large female body size is selected for, which is seen in the family Araneidae. All Argiope species, including Argiope bruennichi, use this method. Some males evolved ornamentation including binding the female with silk, having proportionally longer legs, modifying the female's web, mating while the female is feeding, or providing a nuptial gift in response to sexual cannibalism. Male body size is not under selection due to cannibalism in all spider species such as Nephila pilipes, but is more prominently selected for in less dimorphic species of spiders, which often selects for larger male size.

Fish

There are cases where males are substantially larger than females. An example is *Lamprologus callipterus*, a type of cichlid fish. In this fish, the males are characterized as being up to 60 times larger than the females. The male's increased size is believed to be advantageous because males collect and defend empty snail shells in each of which a female breeds. Males must be larger and more powerful in order to collect the largest shells. The female's body size must remain small because in order for her to breed, she must lay her eggs inside the empty shells. If she grows too large, she will not fit in the shells and will be unable to breed. Another example is the dragonet, in which males are considerably larger than females and possess longer fins.

The female's small body size is also likely beneficial to her chances of finding an unoccupied shell. Larger shells, although preferred by females, are often limited in availability. Hence, the female is limited to the growth of the size of the shell and may actually change her growth rate according to shell size availability. In other words, the male's ability to collect large shells depends on his size. The larger the male, the larger the shells he is able to collect. This then allows for females to be larger in his brooding nest which makes the difference between the sizes of the sexes less substantial. Male-male competition in this fish species also selects for large size in males. There is aggressive competition by males over territory and access to larger shells. Large males win fights and steal shells from competitors. Sexual dimorphism also occurs in hermaphroditic fish. These species are known as sequential hermaphrodites. In fish, reproductive histories often include the sex-change from female to male where there is a strong connection between growth, the sex of an individual, and the mating system it operates within. In protogynous mating systems where males dominate mating with many females, size plays a significant role in male reproductive success. Males have a propensity to be larger than females of a comparable age but it is unclear whether the size increase is due to a growth spurt at the time of the sexual transition or due to the history of faster growth in sex changing individuals. Larger males are able to stifle the growth of females and control environmental resources.

Social organization plays a large role in the changing of sex by the fish. It is often seen that a fish will change its sex when there is a lack of dominant male within the social hierarchy. The females that change sex are often those who attain and preserve an initial size advantage early in life. In either case, females which change sex to males are larger and often prove to be a good example of dimorphism.

In other cases with fish, males will go through noticeable changes in body size, and females will go through morphological changes that can only be seen inside of the body. For example, in sockeye salmon, males develop larger body size at maturity, including an increase in body depth, hump height, and snout length. Females experience minor changes in snout length, but the most noticeable difference is the huge increase in gonad size, which accounts for about 25% of body mass.

Sexual selection was observed for female ornamentation in Gobiusculus flavescens, known as two-spotted gobies. Traditional hypotheses suggest that male-male competition drives selection. However, selection for ornamentation within this species suggests that showy female traits can be selected through either female-female competition or male mate choice. Since carotenoid-based ornamentation suggests mate quality, female two-spotted guppies that develop colorful orange bellies during the breeding season are considered favorable to males. The males invest heavily in offspring during the incubation, which leads to the sexual preference in colorful females due to higher egg quality.

Amphibians and Reptiles

In amphibians and reptiles, the degree of sexual dimorphism varies widely among taxonomic groups. The sexual dimorphism in amphibians and reptiles may be reflected in any of the following: anatomy; relative length of tail; relative size of head; overall size as in many species of vipers and lizards; coloration as in many amphibians, snakes, and lizards, as well as in some turtles; an ornament as in many newts and lizards; the presence of specific sex-related behaviour is common to many lizards; and vocal qualities which are frequently observed in frogs.

Anolis lizards show prominent size dimorphism with males typically being significantly larger than females. For instance, the average male *Anolis sagrei* was 53.4 mm vs. 40 mm in females. Different sizes of the heads in anoles have been explained by differences in the estrogen pathway. The sexual dimorphism in lizards is generally attributed to the effects of sexual selection, but other mechanisms including ecological divergence and fecundity selection provide alternative explanations. The development of color dimorphism in lizards is induced by hormonal changes at the onset of sexual maturity, as seen in Psamodromus algirus, Sceloporus gadoviae, and S. undulates erythrocheilus.

Birds

Mandarin ducks, male (left) and female (right)

Sexual dimorphism in birds can be manifested in size or plumage differences between the sexes. Sexual size dimorphism varies among taxa with males typically being larger, though this is not always the case i.e. birds of prey, hummingbirds, and some species of flightless birds. Plumage dimorphism, in the form of ornamentation or coloration, also varies, though males are typically the more ornamented or brightly colored sex. Such differences have been attributed to the unequal reproductive contributions of the sexes. This difference produces a stronger female choice since they have more risk in producing offspring. In some species, the male's contribution to reproduction ends at copulation, while in other species the male becomes the main caregiver. Plumage polymorphisms have evolved to reflect these differences and other measures of reproductive fitness, such as body condition or survival. The male phenotype sends signals to females who then choose the 'fittest' available male.

Sexual dimorphism is a product of both genetics and environmental factors. An example of sexual polymorphism determined by environmental conditions exists in the red-backed fairywren. Red-backed fairywren males can be classified into three categories during breeding season: black breeders, brown breeders, and brown auxiliaries. These differences arise in response to the bird's body condition: if they are healthy they will produce more androgens thus becoming black breeders, while less healthy birds produce less androgens and become brown auxiliaries. The reproductive success of the male is thus determined by his success during each year's non-breeding season, causing reproductive success to vary with each year's environmental conditions.

Migratory patterns and behaviors also influence sexual dimorphisms. This aspect also stems back to the size dimorphism in species. It has been shown that the larger males are better at coping with the difficulties of migration and thusly are more successful in reproducing when reaching the breeding destination. When viewing this in an evolutionary standpoint many theories and explanations come to consideration. If these are the result for every migration and breeding season the expected results should be a shift towards a larger male population through sexual selection. Sexual selection is strong when the factor of environmental selection is also introduced. The environmental selection may support a smaller chick size if those chicks were born in an area that allowed them to grow to a larger size, even though under normal conditions they would not be able to reach this optimal size for migration. When the environment gives advantages and disadvantages of this sort, the strength of selection is weakened and the environmental forces are given greater morphological weight. The sexual dimorphism could also produce a change in timing of migration leading to differences in mating success within the bird population. When the dimorphism produces that large of a variation between the sexes and between the members of the sexes multiple evolutionary effects can take place. This timing could even lead to a speciation phenomenon if the variation becomes strongly drastic and favorable towards two different outcomes.

Sexual dimorphism is maintained by the counteracting pressures of natural selection and sexual selection. For example, sexual dimorphism in coloration increases the vulnerability of bird species to predation by European sparrowhawks in Denmark. Presumably, increased sexual dimorphism means males are brighter and more conspicuous, leading to increased predation. Moreover, the production of more exaggerated ornaments in males may come at the cost of suppressed immune function. So long as the reproductive benefits of the trait due to sexual selection are greater than the costs imposed by natural selection, then the trait will propagate throughout the population. Reproductive benefits arise in the form of a larger number of offspring, while natural selection imposes costs in the form of reduced survival. This means that even if the trait causes males to die earlier, the trait is still beneficial so long as males with the trait produce more offspring than males lacking the trait. This balance keeps the dimorphism alive in these species and ensures that the next generation of successful males will also display these traits that are attractive to the females.

Such differences in form and reproductive roles often cause differences in behavior. As previously stated, males and females often have different roles in reproduction. The courtship and mating behavior of males and females are regulated largely by hormones throughout a bird's lifetime. Activational hormones occur during puberty and adulthood and serve to 'activate' certain behaviors when appropriate, such as territoriality during breeding season. Organizational hormones occur only during a critical period early in development, either just before or just after hatching in most birds, and determine patterns of behavior for the rest of the bird's life. Such behavioral differences can cause disproportionate sensitivities to anthropogenic pressures. Females of the whinchat in Switzerland breed in intensely managed grasslands. Earlier harvesting of the grasses during the breeding season lead to more female deaths. Populations of many birds are often male-skewed and when sexual differences in behavior increase this ratio, populations decline at a more rapid rate. Also not all male dimorphic traits are due to hormones like testosterone, instead they are a naturally occurring part of development, for example plumage.

Sexual dimorphism may also influence differences in parental investment during times of food scarcity. For example, in the blue-footed booby, the female chicks grow faster than the males, resulting in booby parents producing the smaller sex, the males, during times of food shortage. This then results in the maximization of parental lifetime reproductive success.

Sexual dimorphism may also only appear during mating season, some species of birds only show dimorphic traits in seasonal variation. The males of these species will molt into a less bright or less exaggerated color during the off breeding season. This occurs because the species is more focused on survival than reproduction, causing a shift into a less ornate state.

Consequently, sexual dimorphism has important ramifications for conservation. However, sexual dimorphism is not only found in birds and is thus important to the conservation of many animals. Such differences in form and behavior can lead to sexual segregation, defined as sex differences in space and resource use. Most sexual segregation research has been done on ungulates, but such research extends to bats, kangaroos, and birds. Sex-specific conservation plans have even been suggested for species with pronounced sexual segregation.

The term sesquimorphism (the Latin numeral prefix *sesqui*- means one-and-one-half, so halfway between *mono*- (one) and *di*- (two)) has been proposed for bird species in which "both sexes have basically the same plumage pattern, though the female is clearly distinguishable by reason of her paler or washed-out colour". Examples include Cape sparrow (*Passer melanurus*), rufous sparrow (subspecies *P. motinensis motinensis*), and saxaul sparrow (*P. ammodendri*).

Mammals

Just like in birds, the brains of many mammals, including humans, are significantly different for males and females of the species. Both genes and hormones affect the formation of many animal brains before "birth" (or hatching), and also behaviour of adult individuals. Hormones significantly affect human brain formation, and also brain development at puberty. A 2004 review in *Nature Reviews Neuroscience* observed that "because it is easier to manipulate hormone levels than the expression of sex chromosome genes, the effects of hormones have been studied much more extensively, and are much better understood, than the direct actions in the brain of sex chromosome genes." It concluded that while "the differentiating effects of gonadal secretions seem to be dominant," the existing body of research "support the idea that sex differences in neural expression of X and Y genes significantly contribute to sex differences in brain functions and disease."

Pinnipeds

Marine mammals show some of the greatest sexual size differences of mammals. The mating system of pinnipeds varies from polygyny to serial monogamy. Pinnipeds are known for early differential growth and maternal investment since the only nutrients for newborn pups is the milk provided by the mother. For example, the males are significantly larger than the females at birth in sea lion pups. The pattern of differential investment can be varied principally prenatally and post-natally. *Mirounga leonina*, the southern elephant seal, is one of the most dimorphic mammals.

Sexual dimorphism in elephant seals is associated with the ability of a male to defend territories, which correlates with polygynic behavior. The large sexual size dimorphism is due to sexual selection, but also because females reach reproductive age much earlier than males. In addition the males do not provide parental care for the young and allocate more energy to growth. This is supported by the secondary growth spurt in males during adolescent years.

Primates

Humans

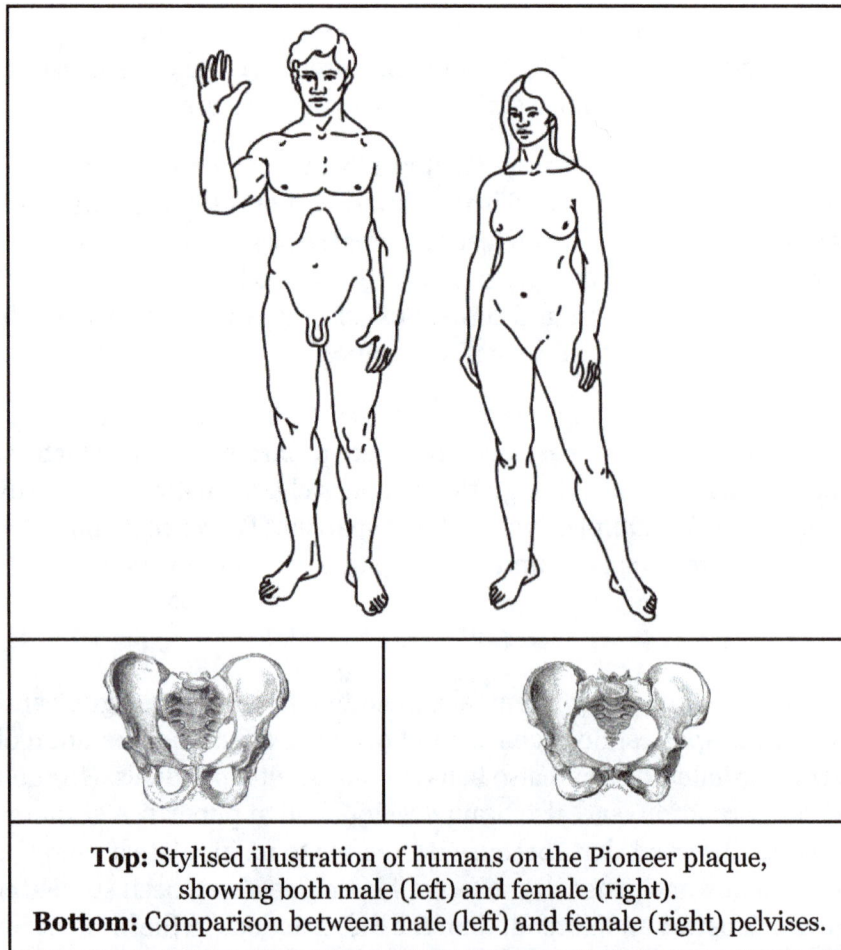

Top: Stylised illustration of humans on the Pioneer plaque, showing both male (left) and female (right).
Bottom: Comparison between male (left) and female (right) pelvises.

In humans, biological sex is determined by five factors present at birth: the presence or absence of a Y chromosome, the type of gonads, the sex hormones, the internal reproductive anatomy (such as the uterus in females), and the external genitalia. Generally, the five factors are either all male or all female. Sexual ambiguity is rare in humans, but wherein such ambiguity does occur, the individual is biologically classified as intersex.

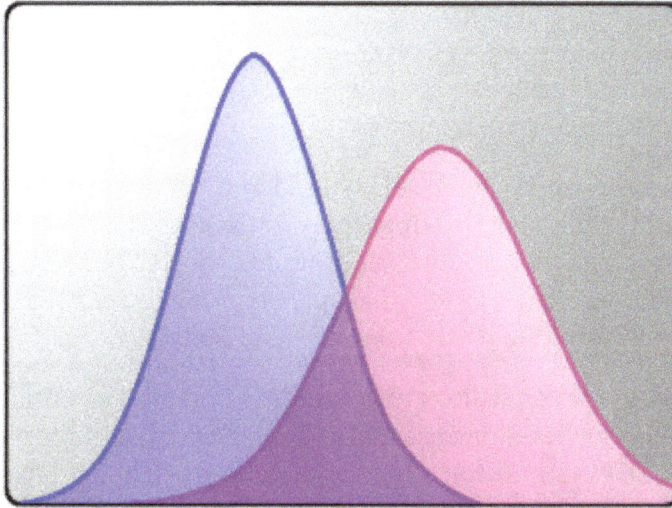

This is a remake of a facial geometric sexual dimorphism diagram from Valenzano, D. R. et al. (2006). The blue bell curve on the left represents the male faces, and the pink bell curve on the right represents the female faces. The purple area in the center represents the overlap of the two bell curves where the feminine male faces cannot be distinguished from the masculine female faces. The bell curves show that the proportion of female faces that are more feminine than the most feminine male faces is much greater than the proportion of male faces that are more masculine than the most masculine female faces.

Sexual dimorphism among humans includes differentiation among gonads, internal genitals, external genitals, breasts, muscle mass, height, the endocrine (hormonal) systems and their physiological and behavioral effects. Human sexual differentiation is effected primarily at the gene level, by the presence or absence of a Y-chromosome, which encodes biochemical modifiers for sexual development in males. According to Clark Spencer Larsen, modern day *Homo sapiens* show a range of sexual dimorphism, with average body mass difference between the sexes being roughly equal to 15% . Ever since Charles Darwin's *The Descent of Man and Selection in Relation to Sex* in 1871 was published, there's been controversy regarding the social, cultural, and political significance of human sexual dimorphism.

The average basal metabolic rate is about 6 percent higher in adolescent males than females and increases to about 10 percent higher after puberty. Females tend to convert more food into fat, while males convert more into muscle and expendable circulating energy reserves. Aggregated data of absolute strength indicates that females have, on average, 40-60% the upper body strength of males, and 70-75% the lower body strength. The difference in strength relative to body mass is less pronounced in trained individuals. In Olympic weightlifting, male records vary from 5.5× body mass in the lowest weight category to 4.2× in the highest weight category, while female records vary from 4.4× to 3.8×, a weight adjusted difference of only 10-20%, and an absolute difference of about 30% (i.e. 472 kg vs 333 kg for unlimited weight classes). A study, carried about by analyzing annual world rankings from 1980–1996, found that males' running times were, on average, 11% faster than females'.

Females are taller, on average, than males in early adolescence, but males, on average, surpass them in height in later adolescence and adulthood. In the United States, adult males are, on average, 9% taller and 16.5% heavier than adult females.

Males typically have larger tracheae and branching bronchi, with about 30 percent greater lung volume per body mass. On average, males have larger hearts, 10 percent higher red blood cell count, higher hemoglobin, hence greater oxygen-carrying capacity. They also have higher circulating clotting factors (vitamin K, prothrombin and platelets). These differences lead to faster healing of wounds and higher peripheral pain tolerance.

Females typically have more white blood cells (stored and circulating), more granulocytes and B and T lymphocytes. Additionally, they produce more antibodies at a faster rate than males. Hence they develop fewer infectious diseases and succumb for shorter periods. Ethologists argue that females, interacting with other females and multiple offspring in social groups, have experienced such traits as a selective advantage.

Considerable discussion in academic literature concerns potential evolutionary advantages associated with sexual competition (both intrasexual and intersexual) and short- and long-term sexual strategies. According to Daly and Wilson, "The sexes differ more in human beings than in monogamous mammals, but much less than in extremely polygamous mammals." One proposed explanation is that human sexuality has developed more in common with its close relative the bonobo, who have similar sexual dimorphism and which are polygynandrous and use recreational sex to reinforce social bonds and reduce aggression.

In the human brain, a difference between sexes was observed in the transcription of the PCDH11X-/Y gene pair unique to *Homo sapiens*. Sexual differentiation in the human brain from the default female state is triggered by testosterone from the fetal testis. Testosterone is converted to estrogen in the brain through the action of the enzyme aromatase. Testosterone acts on many brain areas, including the SDN-POA, to create the masculinized brain pattern. Female brains may be shielded from the masculinizing effects of estrogen through the action of sex-hormone binding globulin and a-fetoprotein.

The relationship between sex differences in the brain and human behavior is a subject of controversy in psychology and society at large. Many females tend to have a higher ratio of gray matter in the left hemisphere of the brain in comparison to males. Males on average have larger brains than females, however when adjusted for total brain volume the gray matter differences between sexes is almost nonexistent. Thus, the percentage of gray matter appears to be more related to brain size than it is to sex. Differences in brain physiology between sexes do not necessarily relate to differences in intellect. Haier *et al.* found in a 2004 study that "men and women apparently achieve similar IQ results with different brain regions, suggesting that there is no singular underlying neuroanatomical structure to general intelligence and that different types of brain designs may manifest equivalent intellectual performance".

Cells

Phenotypic differences between sexes are evident even in cultured cells from tissues. For example, female muscle-derived stem cells have a better muscle regeneration efficiency than male ones. There are reports of several metabolic differences between male and female cells and they also respond to stress differently.

Larger Females are Reproductively Advantageous

In theory, larger females are favored by competition for mates, especially in polygamous species. Larger females offer an advantage in fertility, since the physiological demands of reproduction are limiting in females. Hence there is a theoretical expectation that females tend to be larger in species that are monogamous. Females are larger in many species of insects, many spiders, many fish, many reptiles, owls, birds of prey and certain mammals such as the spotted hyena, and baleen whales such as blue whale. As an example, in some species, females are sedentary and sparsely distributed, and so males must search for them. Fritz Vollrath and Geoff Parker argue that this difference in behaviour leads to radically different selection pressures on the two sexes, evidently favouring smaller males. Cases where the male is larger than the female have been studied as well, and require alternative explanations.

One example of this type of sexual size dimorphism is the bat *Myotis nigricans*, where females are substantially larger than males in terms of body weight, skull measurement, and forearm length. The interaction between the sexes and the energy needed to produce viable offspring make it favorable for females to be larger in this species. Females bear the energetic cost of producing eggs, which is much greater than the cost of making sperm by the males. The fecundity advantage hypothesis states that a larger female is able to produce more offspring and give them more favorable conditions to ensure their survival; this is true for most ectotherms. A larger female can provide parental care for a longer time while the offspring matures. The gestation and lactation periods are fairly long in *M. nigricans*, the females suckling their offspring until they reach nearly adult size. They would not be able to fly and catch prey if they did not compensate for the additional mass of the offspring during this time. Smaller male size may be an adaptation to increase maneuverability and agility, allowing males to compete better with females for food and other resources.

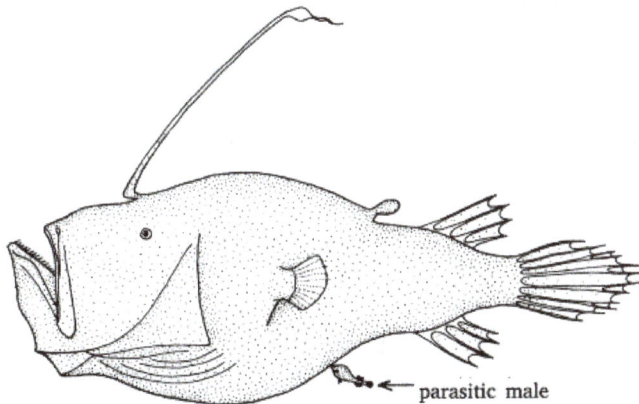

Female triplewart seadevil, an anglerfish, with male attached near vent (arrow)

Some species of anglerfish also display extreme sexual dimorphism. Females are more typical in appearance to other fish, whereas the males are tiny rudimentary creatures with stunted digestive systems. A male must find a female and fuse with her: he then lives parasitically, becoming little more than a sperm-producing body in what amounts to an effectively hermaphrodite composite organism. A similar situation is found in the Zeus water bug *Phoreticovelia disparata* where the female has a glandular area on her back that can serve to feed a male, which clings to her (note that although males can survive away from females, they generally are not free-living).

Some plant species also exhibit dimorphism in which the females are significantly larger than the males, such as in the moss *Dicranum* and the liverwort *Sphaerocarpos*. There is some evidence that, in these genera, the dimorphism may be tied to a sex chromosome, or to chemical signalling from females.

Another complicated example of sexual dimorphism is in the *Vespula squamosa*, or southern yellowjacket. In this wasp species, the female workers are the smallest, the male workers are slightly larger, and the female queens are significantly larger than her female worker and male counterparts.

Evolution

Sexual dimorphism by size is evident in some extinct species such as the velociraptor. In the case of velociraptors the sexual size dimorphism may have been caused by two factors: male competition for hunting ground to attract mates, and/or female competition for nesting locations and mates, males being a scarce breeding resource.

In 1871, Charles Darwin advanced the theory of sexual selection, which related sexual dimorphism with sexual selection.

It has been proposed that the earliest sexual dimorphism is the size differentiation of sperm and eggs (anisogamy), but the evolutionary significance of sexual dimorphism is more complex than that would suggest. Anisogamy and the usually large number of small male gametes relative to the larger female gametes usually lies in the development of strong sperm competition, because small sperm enable organisms to produce a large number of sperm, and make males (or male function of hermaphrodites) more redundant. This intensifies male competition for mates and promotes the evolution of other sexual dimorphim in many species, especially in vertebrates including mammals. However, in some species, the females can be larger than males, irrespective of gametes, and in some species females (usually of species in which males invest a lot in rearing offspring and thus no longer considered as so redundant) compete for mates in ways more usually associated with males.

In many non-monogamous species, the benefit to a male's reproductive fitness of mating with multiple females is large, whereas the benefit to a female's reproductive fitness of mating with multiple males is small or nonexistent. In these species, there is a selection pressure for whatever traits enable a male to have more matings. The male may therefore come to have different traits from the female.

Male (left), offspring, and female (right) Sumatran orangutans.

These traits could be ones that allow him to fight off other males for control of territory or a harem, such as large size or weapons; or they could be traits that females, for whatever reason, prefer in mates. Male-male competition poses no deep theoretical questions but mate choice does.

Females may choose males that appear strong and healthy, thus likely to possess "good alleles" and give rise to healthy offspring. In some species, however, females seem to choose males with traits that do not improve offspring survival rates, and even traits that reduce it (potentially leading to traits like the peacock's tail). Two hypotheses for explaining this fact are the sexy son hypothesis and the handicap principle.

The sexy son hypothesis states that females may initially choose a trait because it improves the survival of their young, but once this preference has become widespread, females must continue to choose the trait, even if it becomes harmful. Those that do not will have sons that are unattractive to most females (since the preference is widespread) and so receive few matings.

The handicap principle states that a male who survives despite possessing some sort of handicap thus proves that the rest of his genes are "good alleles". If males with "bad alleles" could not survive the handicap, females may evolve to choose males with this sort of handicap; the trait is acting as a hard-to-fake signal of fitness.

Sexual Mimicry

Sexual mimicry occurs when one sex mimics the opposite sex in its behavior, appearance, or chemical signalling. It is more commonly seen within invertebrate species, although sexual mimicry is also seen among vertebrates such as spotted hyenas. Sexual mimicry is commonly used as a mating strategy to gain access to a mate, a defense mechanism to avoid more dominant individuals, or a survival strategy. It can also be a physical characteristic that establishes an individual's place in society. Sexual mimicry is employed differently across species and it is part of their strategy for survival and reproduction. Examples of sexual mimicry in animals include the spotted hyena, certain types of fish, passerine birds and some species of insect among others. These are cases of intraspecific sexual mimicry, but interspecific sexual mimicry can also occur in some plant species, especially orchids. In plants employing sexual mimicry, flowers mimic mating signals of their pollinator insects. These insects are attracted and pollinate the flowers through pseudocopulations or other sexual behaviors performed on the flower.

Social Systems

Sexual mimicry can influence the species' social system. The most common example is the spotted hyenas, *Crocuta crocuta*. Female hyenas resemble male hyenas in their sexual anatomy: the females have peniform clitorises, resembling a penis, and false scrotal sacs. These characteristics, as well as high androgen levels in their blood, make for aggressive females, which results in their dominance over males; the female with the lowest rank is more dominant than the highest-ranking male. Within the female population in each clan, there are different ranks: the dominant females, who reproduce at an earlier age and get more access to food, and the non-dominant females. Their dominance is hierarchical and is passed from mother to daughter. By contrast, male spotted hyenas gain their social status with the length of their stay in the clan; it does not involve aggressive contests. The males leave their clan between the ages of two and six and join a different clan where they gain status with age. Males also foster amicable relationships with the females to stabilize their position in the social hierarchy.

Hyena greeting ceremony

Because females are the dominant sex among spotted hyenas, they are the most respected. Subordinate female hyenas initiate a 'greeting' with dominant female hyenas as a sign of respect and are forced to do so if they refuse. This greeting used by hyenas reflects the asymmetry of their ranking; the animal being greeted (the subordinate individual) extends its hind legs and the individual doing the greeting (the dominant hyena) licks or sniffs the erect peniform clitoris. By lifting its hind legs, the hyena being greeted (the subordinate hyena) exposes its most vulnerable body part to the other individual, an act that reflects inferiority. As well, when its hind legs are lifted, a scent can be identified by the other individual. Subordinate hyenas expose their scent more often than high-ranking hyenas. This greeting, however, is not commonly seen between males and adult females; when it does occur, it is restricted to males of median or higher rank greeting dominant females.

Mating Systems

In the spotted hyenas, the only way for the males to mate with the females is if they have the female's full cooperation because of the female's peniform clitoris. An increase in the male's status gave them more access to dominant females in the clan. Female dominant hyenas do not mate with multiple males, possibly due to the cost of cleaning their genitalia, which hyenas are seen doing after copulation. Because they will get access to the most dominant and better fit males, they do not need to copulate with multiple males to produce offspring of higher fitness. Non-dominant females are observed copulating more often with lower-ranking males. It is costly for female hyenas to give birth through their long peniform clitoris. The umbilical cord is 12–18 cm long, while the journey from the uterus to the clitoris end is 40 cm. The umbilical cord often breaks before the cub emerges, leading to death by anoxia for many young. This journey is not only harmful for the cubs, but also for the mother. The tissue of the clitoris will sometimes rip open when giving birth for the first time which can be fatal to the mother.

Female spotted hyenas are the choosy sex because they invest in parental care as well as being the dominant sex in the clan. However, males are likely to still have a preference for a particular female as it is seen in other animals; high-ranking females begin breeding at a younger age and their offspring are more likely to survive to adulthood than the offspring of low-ranking females. Males associate more closely with females that are fertile, a state most likely noticed through olfactory cues. While middle/high-ranking males associate with high-ranking females, low-ranking males associate equally with high and low-ranking females. Associating with low-ranking females may be due to low-ranking males failing to recognize the reproductive success of high-ranking females or using a different type of reproductive strategy. Males tend to spend lot of time with the female they mate with before conception to avoid other males coming in close contact with her.

Sexual mimicry is also used as a mate-guarding strategy by some species. Mate-guarding is a process in which a member of a species prevents another member of the same species from mating with their partner. Mate-guarding is seen in *Cotesia rubecula*, a parasitic wasp from the Braconidae family whose mating system is polygynous. Males are attracted to females through pheromones and they induce females to mate through vibrations, to which the female responds by assuming a specific position. When a male who has copulated with a female sees another male trying to court her, he will often adopt the female receptive position. Post-copulatory female mimicry by the male offers an advantage by acting as a mate-guarding mechanism. If a second male arrives soon enough after the female copulates with the first male, the second male may be able to induce a second copulation which will compete with the first one. However, if the first male who copulated with her mimics the female, it distracts the second male long enough that the female becomes unreceptive.

Sneaky Copulation

Salaria pavo female

Salaria pavo male

Sneaky copulation is a strategy used by many aquatic organisms who portray sexual mimicry. Several studies have found that small male fish will look and behave like the female of their species in order to gain access to female territory and copulate with them. In the Blenniidae family, the *Salaria pavo*, female bleniid fish will show a specific colour pattern and movement when they want to approach a male and copulate with him. The male guards a territory, and when the female lays her eggs, the parental male protects that territory until the eggs hatch. A second type of males, the sneaker males, is parasitic and resembles the female bleniid fish in their small size, colour, and movement patterns. This allows them to intrude into the nest guarded by the parental males. Sneaker males approach the nests with the same colour patterns and movements that the females hold. Most cases of sneaker males are seen when there is a female already inside the nest although sometimes the sneaker fish enters the nest alongside a female. This species of fish releases the sperm before the female releases her eggs into the water making it possible for the sneaker fish to fertilize an egg, even if the female is not present in the nest.

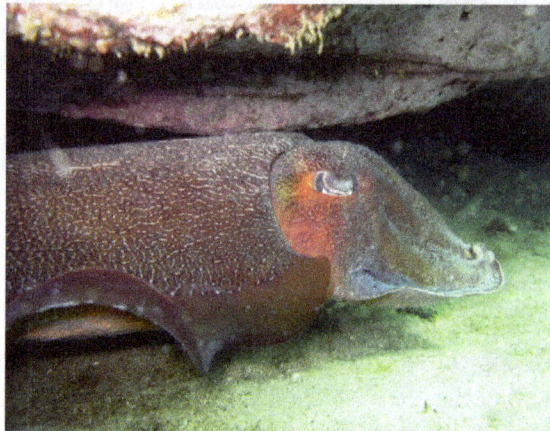

Sepia apama

In the Sepiina family, *Sepia apama*, also known as cuttlefish, have some males that are large and able to guard a female's nest while other males are small and resemble females in order to sneak in copulations. In the giant cuttlefish, the male courts the female and transfers its sperm to a pouch below the female's beak. During this process, the female displays a body pattern of black splotches on a white background. Once the eggs are laid, the male guards the nest from any possible suitors and opponents. A 'second female' is sometimes seen during male-female interaction in close proximity to the couple. This female-looking cuttlefish has the same black blotches as a real female. If the male leaves to fight other males, this individual approaches the female and copulates with her, usually with success. However, in the absence of rivals, these 'mimicking female' males display the phenotype of a mature male.

Sexual Mimicry against Aggression

A similar phenomenon to the sneaker fish males is observed in the dark-edged splitfin, *Girardinichthys multiradiatus*. The juveniles resemble the pregnant females in the species by having a dark spot near the vent. In this case, however, the mimicking males have the capability to resemble the females or become a morphologically mature male throughout most of their adult life. This dark spot allows the female-looking males to escape aggression from more dominant males, as well as reducing the chance of having a female nearby flee due to persisting courting males. The mature males do not attack the subordinate fish and the subordinate fish decides when to initiate the fights, which gives it an advantage as the mature male is not expecting this. The dark spot also permits access of subordinate males to females, a characteristic that is advantageous because females' eggs can only be fertilized during a five-day fertilizing window.

Male *Ficedula hypoleuca*

Female/juvenile *Ficedula hypoleuca*

Sexual mimicry to avoid aggression is also seen in birds. In some bird species, males have a female-like plumage colour during their second year of life (SY males). These SY males are sexually mature and able to breed, but their morphology differs greatly from the older, after second year (ASY) males. Various studies have looked into this delayed plumage maturation (DPM) and found that the DPM in SY males reduces aggression from ASY males. Female mimicry in birds was first found in European-pied flycatcher, *Ficedula hypoleuca*. When a dull-coloured male is in the area, mature males reduce their aggressiveness and behave as if the intruder is a female. The dull plumage is seen mostly in younger males, likely due to being born later in the previous spring. The resemblance to females benefit these young males when trying to occupy a territory with many males already present because the young males can gain information and access to a territory that would not be accessible to them otherwise.

There is a big cost to not looking like a male when it comes to defending a territory or attracting a mate. Females show aggression against dull-coloured males, making it harder for them to mate. However, DPM has some benefits: as mentioned above, it reduces aggression from older males. As well, these female-looking birds are able to get access to territories, mates, and food that may be not be available to them otherwise. Another benefit is that DPM provides SY birds with a longer lifespan; because they do not have to compete with other males, their mortality rate is lower. This advantage, however, only benefits individuals of species that have a longer potential lifespan and, therefore, DPM would not benefit a short-lived species. This is known as the breeding threshold hypothesis, and states that SY males should only delay breeding if there is a large mortality difference between the SY males who attempt to breed and the ones who do not.

Platysaurus broadleyi

Most studies addressed DPM as a type of sexual mimicry, which is done through deception: male ASY birds should not be able to tell females or SY males apart. However, Muheter et al. (1997) found that territorial males perceive the dull-coloured males as males but they show less aggression because their dull-coloured plumage promotes low competitive ability. They referred to this as honest signalling and not sexual mimicry.

Another example of sexual mimicry occurs in Broadley's Flat Lizard, *Platysaurus broadleyi*, where some males mimic females. Flat lizard males tend to be territorial and aggressive towards other males. Therefore, it is beneficial for some males to mimic females in order to avoid aggressive encounters and move freely through the male's territory, looking for mates. There are two types of males in this population; she-males, who mimic females, and he-males, who look like males. The she-males can visually fool the he-males into believing that they are female due to their female morphology. However, the she-males cannot fool the he-males through scent, as he-males can detect the difference. Therefore, the most successful she-males are those who avoid close contact with other males, thereby reducing the chances of detection through chemical signals.

Molecular Control over Sexual Mimicry

Female hyenas' sexual mimicry to males is part of their anatomy and it is thought to have evolved through high androgen levels. While female ancestors were smaller than males, selection must have acted upon androgen levels and female body size to increase both, leading to further selection and larger females than males. The high androgen levels are not present in the female ovaries, as it was once thought; the stromal tissue in the ovaries contains lower testosterone levels than the males' testes. However, females' androgen levels in the blood are as high as the ones found in the male, having the effect of morphologically male-looking females.

Ruffs can also show sexual mimicry through a combination of genetics and hormones. In a population of ruffs, *Philomachus pugnax*, there are three types of male morphs: independent males and satellite males, both of which are reproductive competitors, and faeder ruffs that resemble females in their plumage. The first two morphs are controlled by a dominant allele at a single auto-somal locus, while the third morph is likely to have come from a combination of a third allele and a lack of testosterone. When testosterone is administered to reeves (female ruffs), male courtship behaviour and male feather colouration are expressed in the reeves. Testosterone, in this case, expresses sex-limited characteristics by acting on the single autosomal gene. Similarly, while it has not yet been tested, it is likely that the lack of testosterone is the cause for the faeder ruffs' similarity to females.

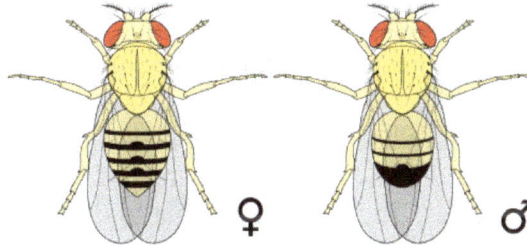

Biology Illustration Animals Insects *Drosophila melanogaster*

A different example is seen in mature female fruit flies, *Drosophila melanogaster*, who are very attractive but their level of attractiveness decreases by half or more after three minutes of mating. Males release a compound, 7-tricosene, into the female during courtship that lowers female attrac-tiveness. However, the researcher found that the females release this compound as well, six hours after mating. This compound lowers the female's levels of attractiveness both times, when the male is courting her and during mating. This way, the female mimics the male and with this compound, she lowers her levels of attractiveness .

Genetic Control Over Sexual Mimicry

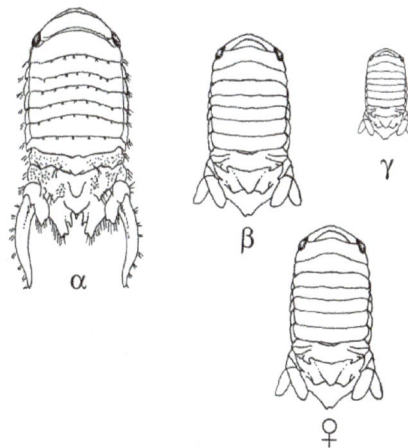

Paracerceis sculpta (Marine isopod)

Some organisms' sexual mimicry is genetically determined by specific alleles. Unlike sexual mim-icry that arises due to molecular compounds or hormones and can sometimes be induced through these molecules, this sexual mimicry arises from the organism's genetic material. Besides the fe-male hyenas' sexual anatomy, which is part of their genetics, some other organisms have only some males/females in their population who look like the opposite sex and this is determined by specific alleles.

In the marine isopod population, *Paracerceis sculpta*, there are three different male morphologies: the alpha male is the largest morph, it matures last, and it is the one who gets privileged access to the females. The beta male is of intermediate size, and it mimics the female to get access to females. Last, the gamma male is the smallest morph and it invades harems, where females go to mate with alpha males, for mating opportunities. This morphology is associated with a single autosomal gene and three different alleles. Beta is the most dominant allele, followed by gamma, which is followed by alpha. Selection on these alleles acts according to the Hardy-Weinberg equilibrium and mating success is equivalent among all three morphs.

The alpha males, who are homozygous for the alpha allele, mate with many females in a harem. The females prefer to aggregate with other females in the harem, which gives the alpha male a bigger selection of mating partners. Shuster (1992) looked at the behaviour and relationship of each morph with respect to the harem and found that beta and gamma males could locate harems that have sexually receptive females. They were also able to differentiate between a harem with a sexually receptive female, i.e. one that is able to mate, and a non-sexually receptive female, i.e. one that has already deposited the embryo into her pouch and can no longer mate. While it is still unclear how the beta males do this or how their mating strategies work, they are not harassed by alpha males due to their mimicry of females: the beta males can attract other females into the harem since females like to go where other females are, and this provides the alpha males with more mates.

Another order of organisms whose sexual mimicry is influenced by their DNA is the Odonata, carnivorous insects known as dragonflies and damselflies. In these species, it is the female who sometimes mimics the male. Within a species, groups of females will differ in colour: one group mimics the males' colour and they are known as androchromes. Other groups will have their own female colouration and they are known as gynochromes. In *Ischnura elegans*, androchromes comprise 6-30% of the female population and their colour is usually blue, like the males; in some populations, androchromes are larger in size than gynochromes. This polymorphism is controlled by an autosomal allele and some studies have looked at the reason for the polymorphism's maintenance.

Aeshna affinis Male

Aeshna affinis Female Androchrome

The most likely theory for the maintenance of the polymorphism in Odonata is the density dependence theory that states that at a high male density, the androchromes are not bothered by the males and their existence is not threatened by male harassment. This hypothesis also assumes that males cannot distinguish between androchromes and other males. This advantage, however, is counteracted with the fact that they will not get a lot of mating opportunities (if any) and their reproduction is limited. This theory is the most likely explanation for the maintenance of polymorphism, since studies have shown that there is an advantage for androchromes in high male-density populations.

Self-control over Sexual Mimicry

While, as seen before, most organisms which portray sexual mimicry are born with this morphology/behaviour, this is not always the case. The giant cuttlefish, *Sepia apama*, mentioned above in the section "sneaky copulations", is born with the capacity to choose whether to change its morphology to look like a female or a mature male. When no competition is seen nearby, the cuttlefish will look like a mature male and mate with the female. However, when a mature male and a female are copulating, the giant cuttlefish will resemble a female and stay at a close distance of the couple, hoping for a chance to mate with the female if the mature male leaves to fight other males. Another example of an organism that has the capability to remain small and look like a female, or become a morphologically mature male, is the dark-edged splitfin, *Girardinichthys multiradiatus*. The purpose for their female mimicry was seen before, in the "sexual mimicry against aggression" section where the female-looking males will escape aggression from dominant males and avoid females fleeing their company due to persisting courting males.

Interspecific Deceptive Mimicry

Interspecific sexual mimicry can also occur in some plant species. The most common example of this is known as sexually deceptive pollination and is found among some orchids. The orchid mimics its pollinator's females, usually hymenopterans such as wasps and bees, attracting the males to the flower. Orchid flowers mimic the sex pheromones and to some degree the visual appearance of the female insect of its pollinator species. The primacy of olfactory over visual cues has been demonstrated in many cases, such as in the European orchid genus Ophrys as well as many Australian sexually deceptive orchids.

Bee Orchid

In few other cases, such as the South African daisy *Gorteria diffusa*, visual signals seem to be of primary importance. Visual signals also enhance the attractiveness of the flowers of some Ophrys species to their pollinators. Some male scoliid wasps such as *Campsoscolia ciliata* are more attracted to the Ophrys flowers' odours than to the odours of the female wasps, although they both attract the males with the same compounds. This is most likely a result of a higher amount of scent coming from the orchid flowers; female wasps tend to produce less scent to avoid attracting predators. Regardless of whether orchids use appearances, fragrances or both, they mimic the female pollinator for their own benefit.

References

- Michod, RE; Levin, BE, eds. (1987). The Evolution of sex: An examination of current ideas. Sunderland, Massachusetts: Sinauer Associates. ISBN 978-0878934584.

- Lynn Margulis; Heather I. McKhann; Lorraine Olendzenski (2001). Illustrated glossary of protoctista: vocabulary of the algae, apicomplexa, ciliates, foraminifera, microspora, water molds, slime molds, and the other protoctists. Jones & Bartlett learn. ISBN 978-0-86720-081-2.

- James Desmond Smyth; Derek Wakelin (1994). Introduction to animal parasitology (3 ed.). Cambridge University Press. pp. 101–102. ISBN 0-521-42811-4.

- R. S. Mehrotra; K. R. Aneja (December 1990). An Introduction to Mycology. New Age International. pp. 83 ff. ISBN 978-81-224-0089-2. Retrieved 4 August 2010.

- Kathleen M. Cole; Robert G. Sheath (1990). Biology of the red algae. Cambridge University Press. pp. 469–. ISBN 978-0-521-34301-5. Retrieved 4 August 2010.

- Edward G. Reekie; Fakhri A. Bazzaz (28 October 2005). Reproductive allocation in plants. Academic Press. pp. 99–. ISBN 978-0-12-088386-8. Retrieved 4 August 2010.

- Robert W. Goy & Bruce S. McEwen (1980). Sexual Differentiation of the Brain: Based on a Work Session of the Neurosciences Research Program. Boston: MIT Press. ISBN 978-0-262-57207-1.

- Knox, David; Schacht, Caroline. Choices in Relationships: An Introduction to Marriage and the Family. 11 ed. Cengage Learning; 2011-10-10 [cited 17 June 2013]. ISBN 9781111833220. p. 64–66.

- Alfred Glucksman (1981). Sexual Dimorphism in Human and Mammalian Biology and Pathology. Academic

Press. pp. 66–75. ISBN 978-0-12-286960-0. OCLC 7831448.

- Gersh, Eileen S.; Gersh, Isidore (1981). Biology of Women. Baltimore: University Park Press (original from the University of Michigan). ISBN 978-0-8391-1622-6.

- Martin Daly & Margo Wilson (1996). "Evolutionary psychology and marital conflict". In David M. Buss & Neil M. Malamuth. Sex, Power, Conflict: Evolutionary and Feminist Perspectives. Oxford University Press. p. 13. ISBN 978-0-19-510357-1.

- Christopher Ryan & Cacilda Jethá (2010). Sex at Dawn: The Prehistoric Origins of Modern Sexuality. Harper. ISBN 978-0-06-170780-3.

- Cordelia Fine (August 2010). Delusions of Gender: How Our Minds, Society, and Neurosexism Create Difference (1st ed.). W. W. Norton & Company. ISBN 978-0-393-06838-2.

- Rebecca Jordan-Young (September 2010). Brain Storm: The Flaws in the Science of Sex Differences. Harvard University Press. ISBN 978-0-674-05730-2.

- A. Jonathan Shaw (2000). "Population ecology, population genetics, and microevolution". In A. Jonathan Shaw & Bernard Goffinet. Bryophyte Biology. Cambridge: Cambridge University Press. pp. 379–380. ISBN 978-0-521-66097-6.

- Howard A. Crum & Lewis E. Anderson (1980). Mosses of Eastern North America. 1. New York: Columbia University Press. p. 196. ISBN 978-0-231-04516-2.

- Bernstein H, Hopf FA, Michod RE (1987). "The molecular basis of the evolution of sex". Adv. Genet. Advances in Genetics. 24: 323–70. doi:10.1016/s0065-2660(08)60012-7. ISBN 9780120176243.

- Avise, J. (2008) Clonality: The Genetics, Ecology and Evolution of Sexual Abstinence in Vertebrate Animals. Oxford University Press. ISBN 019536967X ISBN 978-0195369670

- Helen Nilsson Sköld; Matthias Obst; Mattias Sköld & Bertil Åkesson (2009). "Stem Cells in Asexual Reproduction of Marine Invertebrates". In Baruch Rinkevich & Valeria Matranga. Stem Cells in Marine Organisms. Springer. p. 125. ISBN 978-90-481-2766-5.

- Neuhof, Moran; Levin, Michael; Rechavi, Oded (26 August 2016). "Vertically and horizontally-transmitted memories – the fading boundaries between regeneration and inheritance in planaria". Biology Open: bio.020149. doi:10.1242/bio.020149. PMID 27565761. Retrieved 30 August 2016.

Diverse Aspects of Zoology

Zoological sciences contain various theories about the natural world. Such theories are derived from phenomena that can be observed among animals in the wild. These theories seek to expand the scope of zoology while being faithful to its basic principles. Some themes explored in this chapter include alpha behaviour among male members of species, bipedalism and the autotomy of limbs.

Phylum

In biology, a phylum is a taxonomic rank below kingdom and above class. Traditionally, in botany the term division was used instead of "phylum", although from 1993 the International Code of Nomenclature for algae, fungi, and plants accepted the designation "phylum". The kingdom Animalia contains approximately 35 phyla, Plantae contains 12, and Fungi contains 7. Current research in phylogenetics is uncovering the relationships between phyla, which are contained in larger clades, like Ecdysozoa and Embryophyta.

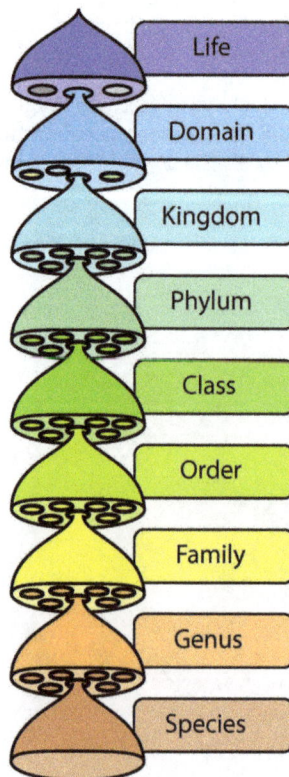

The hierarchy of biological classification's eight major taxonomic ranks. A kingdom contains one or more phyla. Intermediate minor rankings are not shown.

General Description and Familiar Examples

The definitions of zoological phyla have changed importantly from their origins in the six Linnaean classes and the four "embranchements" of Georges Cuvier. Haeckel introduced the term *phylum*, based on the Greek word *phylon* ('tribe' or 'stock'). In plant taxonomy, Eichler (1883) classified plants into five groups, named divisions.

Informally, phyla can be thought of as grouping organisms based on general specialization of body plan. At its most basic, a phylum can be defined in two ways: as a group of organisms with a certain degree of morphological or developmental similarity (the phenetic definition), or a group of organisms with a certain degree of evolutionary relatedness (the phylogenetic definition). Attempting to define a level of the Linnean hierarchy without referring to (evolutionary) relatedness is unsatisfactory, but a phenetic definition is useful when addressing questions of a morphological nature—such as how successful different body plans were.

Definition Based on Genetic Relation

The most important objective measure in the above definitions is the "certain degree"—how unrelated do organisms need to be to be members of different phyla? The minimal requirement is that all organisms in a phylum should be clearly more closely related to one another than to any other group. Even this is problematic because the requirement depends on knowledge of organisms' relationships: as more data become available, particularly from molecular studies, we are better able to judge the relationships between groups. So phyla can be merged or split if it becomes apparent that they are related to one another or not. For example, the bearded worms were described as a new phylum (the Pogonophora) in the middle of the 20th century, but molecular work almost half a century later found them to be a group of annelids, so the phyla were merged (the bearded worms are now an annelid family). On the other hand, the highly parasitic phylum Mesozoa was divided into two phyla, Orthonectida and Rhombozoa, when it was discovered the Orthonectida are probably deuterostomes and the Rhombozoa protostomes.

This changeability of phyla has led some biologists to call for the concept of a phylum to be abandoned in favour of cladistics, a method in which groups are placed on a "family tree" without any formal ranking of group size.

Definition Based on Body Plan

A definition of a phylum based on body plan has been proposed by paleontologists Graham Budd and Sören Jensen (as Haeckel had done a century earlier). The definition was posited because extinct organisms are hardest to classify: they can be offshoots that diverged from a phylum's line before the characters that define the modern phylum were all acquired. By Budd and Jensen's definition, a phylum is defined by a set of characters shared by all its living representatives.

This approach brings some small problems—for instance, ancestral characters common to most members of a phylum may have been lost by some members. Also, this definition is based on an arbitrary point of time: the present. However, as it is character based, it is easy to apply to the fossil record. A greater problem is that it relies on a subjective decision about which groups of organisms should be considered as phyla.

The approach is useful because it makes it easy to classify extinct organisms as "stem groups" to the phyla with which they bear the most resemblance, based only on the taxonomically important similarities. However, proving that a fossil belongs to the crown group of a phylum is difficult, as it must display a character unique to a sub-set of the crown group. Furthermore, organisms in the stem group of a phylum can possess the "body plan" of the phylum without all the characteristics necessary to fall within it. This weakens the idea that each of the phyla represents a distinct body plan.

A classification using this definition may be strongly affected by the chance survival of rare groups, which can make a phylum much more diverse than it would be otherwise. Representatives of many modern phyla did not appear until long *after* the Cambrian.

Known Phyla

Animal Phyla

	Protostome	
	Deuterostome	Bilateria
	Basal/disputed	
	Others (Radiata or Parazoa)	

Phylum	Meaning	Common name	Distinguishing characteristic	Species described
Acanthocephala	Thorny headed worms	Thorny-headed worms	Reversible spiny proboscis that bears many rows of hooked spines	approx. 1,100
Acoelomorpha	Without gut	Acoels	No mouth or alimentary canal (alimentary canal = digestive tract in digestive system)	approx. 350
Annelida	Little ring	Annelids	Multiple circular segment	17,000+ extant
Arthropoda	Jointed foot	Arthropods	Segmented bodies and jointed limbs, with Chitin exoskeleton	1,134,000+
Brachiopoda	Arm foot	Lamp shells	Lophophore and pedicle	300-500 extant
Bryozoa	Moss animals	Moss animals, sea mats	Lophophore, no pedicle, ciliated tentacles, anus outside ring of cilia	5,000 extant
Chaetognatha	Longhair jaw	Arrow worms	Chitinous spines either side of head, fins	approx. 100 extant
Chordata	With a cord	Chordates	Hollow dorsal nerve cord, notochord, pharyngeal slits, endostyle, post-anal tail	approx. 100,000+
Cnidaria	Stinging nettle	Anemones / Jellyfish	Nematocysts (stinging cells)	approx. 11,000
Ctenophora	Comb bearer	Comb jellies	Eight "comb rows" of fused cilia	approx. 100 extant
Cycliophora	Wheel carrying	Symbion	Circular mouth surrounded by small cilia, sac-like bodies	3+

Echinodermata	Spiny skin	Echinoderms	Fivefold radial symmetry in living forms, mesodermal calcified spines	approx. 7,000 extant; approx. 13,000 extinct
Entoprocta	Inside anus	Goblet worm	Anus inside ring of cilia	approx. 150
Gastrotricha	Hair stomach	Hairybacks	Two terminal adhesive tubes	approx. 690
Gnathostomulida	Jaw orifice	Jaw worms		approx. 100
Hemichordata	Half cord	Acorn worms, pterobranchs	Stomochord in collar, pharyngeal slits	approx. 100 extant
Kinorhyncha	Motion snout	Mud dragons	Eleven segments, each with a dorsal plate	approx. 150
Loricifera	Corset bearer	Brush heads	Umbrella-like scales at each end	approx. 122
Micrognathozoa	Tiny jaw animals	—	Accordion-like extensible thorax	1
Mollusca	Soft	Mollusks / molluscs	Muscular foot and mantle round shell	112,000
Nematoda	Thread like	Round worms	Round cross section, keratin cuticle	25,000–1,000,000
Nematomorpha	Thread form	Horsehair worms		approx. 320
Nemertea	A sea nymph	Ribbon worms		approx. 1,200
Onychophora	Claw bearer	Velvet worms	Legs tipped by chitinous claws	approx. 200 extant
Orthonectida	Straight swim		Single layer of ciliated cells surrounding a mass of sex cells	approx. 20
Phoronida	Zeus's mistress	Horseshoe worms	U-shaped gut	11
Placozoa	Plate animals		Differentiated top and bottom surfaces, two ciliated cell layers, amoeboid fiber cells in between	1
Platyhelminthes	Flat worm	Flatworms		approx. 25,000
Porifera*	Pore bearer	Sponges	Perforated interior wall	5,000+ extant
Priapulida	Little Priapus			approx. 16
Rhombozoa	Lozenge animal	—	Single anteroposterior axial cell surrounded by ciliated cells	75
Rotifera	Wheel bearer	Rotifers	Anterior crown of cilia	approx. 2,000
Sipuncula	Small tube	Peanut worms	Mouth surrounded by invertible tentacles	144–320
Tardigrada	Slow step	Water bears	Four segmented body and head	1,000+

Xenacoelomorpha	Strange flatworm	—	Ciliated deuterostome	2
Total: 35				**2,000,000+**

Land plant Phyla (Divisions)

The ten Divisions into which the living embryophytes (land plants) are often placed are shown in the table below. To these may be added two algal Divisions, Chlorophyta and Charophyta, which are included with land plants in the clade Viridiplantae. The definition and classification of plants at this level varies from source to source and has changed progressively in recent years. Thus some sources place horsetails in division Arthrophyta and ferns in division Pteridophyta, while others place them both in Pteridophyta, as shown below. The division Pinophyta may be used for all gymnosperms (i.e. including cycads, ginkgos and gnetophytes), or for conifers alone as below.

Since the first publication of the APG system in 1998, which proposed a classification of angiosperms up to the level of orders, many sources have preferred to treat ranks higher than orders as informal clades. Where formal ranks have been provided, the traditional divisions listed below have been reduced to a very much lower level, e.g. subclasses.

Division	Meaning	Common name	Distinguishing characteristics
Anthocerotophyta	*Anthoceros*-like plants	hornworts	horn-shaped sporophytes, no vascular system
Bryophyta	*Bryum*-like plants, moss plants	mosses	persistent unbranched sporophytes, no vascular system
Marchantiophyta, Hepatophyta	*Marchantia*-like plants, liver plants	liverworts	ephemeral unbranched sporophytes, no vascular system
Lycopodiophyta, Lycophyta	*Lycopodium*-like plants, "wolf" plants	clubmosses & spikemosses	microphyll leaves, vascular system
Pteridophyta	*Pteris*-like plants, fern plants	ferns & horsetails	prothallus gametophytes, vascular system
Pinophyta, Coniferophyta	*Pinus*-like plants, cone-bearing plants	conifers	cones containing seeds and wood composed of tracheids
Cycadophyta	*Cycas*-like plants, palm-like plants	cycads	seeds, crown of compound leaves
Ginkgophyta	*Ginkgo*-like plants	ginkgo, Maidenhair tree	seeds not protected by fruit (single living species)
Gnetophyta	*Gnetum*-like plants	gnetophytes	seeds and woody vascular system with vessels
Magnoliophyta	*Magnolia*-like plants	flowering plants, angiosperms	flowers and fruit, vascular system with vessels
Total: 10			

Fungal Divisions

1. Chytridiomycota
2. Blastocladiomycota
3. Zygomycota
4. Glomeromycota
5. Ascomycota
6. Basidiomycota
7. Microsporidia
8. Neocallimastigomycota

Protista Phyla (Divisions)

Group Description	Phylum	Meaning	Common name	Distinguishing characteristics	Example
Heterotrophs (no locomotor apparatus)	Rhizopoda	root-foot	Amoeba	Amoeboids have the ability to change their shape.	Amoeba
	Actinopoda	ray-foot	-------	long, thin axopodia	Radiolarians
	Foraminifera	hole bearers	Forams	Complex shells with one or more chambers	Forams
Photosynthetic protists	Dinoflagellata	whirling scourge	Dinoflagellates	Unicellular, have two dissimilar flagella	Red tides
	Euglenophyta	Good-eyed plant	Euglenids	Have a pellicle, which gives shape to the cell.	Euglena
	Chrysophyta	golden plant	golden algae		Diatoma
	Rhodophyta	rose plant	red algae	Cells do not have flagella or centrioles; use phycobiliproteins which gives red tint	Coralline Algae
	Phaeophyta	gray plant	brown algae	chloroplasts surrounded by four membranes - form differentiated tissues	Kelp
Heterotrophs (flagella)	Sarcomastigophora				Trypanosoma cruzi
	Ciliophora				Paramecium
Non-motile spore-formers	Apicomplexa				Plasmodium
Heterotrophs (restricted mobility)	Oomycota				Water Molds
	Acrasiomycota				Dictyostelium
	Myxomycota				Fuligo
Total: 14					

Bacterial Phyla/Divisions

Currently there are 29 phyla accepted by List of Prokaryotic names with Standing in Nomenclature (LPSN)

1. Acidobacteria, phenotipically diverse and mostly uncultured

2. Actinobacteria, High-G+C Gram positive species

3. Aquificae, only 14 thermophilic genera, deep branching

4. Bacteroidetes

5. Caldiserica, formerly candidate division OP5, *Caldisericum exile* is the sole representative

6. Chlamydiae, only 6 genera

7. Chlorobi, only 7 genera, green sulphur bacteria

8. Chloroflexi, green non-sulphur bacteria

9. Chrysiogenetes, only 3 genera (*Chrysiogenes arsenatis, Desulfurispira natronophila, Desulfurispirillum alkaliphilum*)

10. Cyanobacteria, also known as the blue-green algae

11. Deferribacteres

12. Deinococcus-Thermus, *Deinococcus radiodurans* and *Thermus aquaticus* are "commonly known" species of this phyla

13. Dictyoglomi

14. Elusimicrobia, formerly candidate division Thermite Group 1

15. Fibrobacteres

16. Firmicutes, Low-G+C Gram positive species, such as the spore-formers Bacilli (aerobic) and Clostridia (anaerobic)

17. Fusobacteria

18. Gemmatimonadetes

19. Lentisphaerae, formerly clade VadinBE97

20. Nitrospira

21. Planctomycetes

22. Proteobacteria, the most known phyla, containing species such as *Escherichia coli* or *Pseudomonas aeruginosa*

23. Spirochaetes, species include *Borrelia burgdorferi*, which causes Lyme disease

24. Synergistetes

25. Tenericutes, alternatively class Mollicutes in phylum Firmicutes (notable genus: *Myco-*

plasma)

26. Thermodesulfobacteria

27. Thermomicrobia

28. Thermotogae, deep branching

29. Verrucomicrobia

Archaeal Phyla/Division/Kingdoms

1. Crenarchaeota, Second most common archaeal phylum

2. Euryarchaeota, most common archaeal phylum

3. Korarchaeota

4. Nanoarchaeota, ultra-small symbiotes, single known species

5. Thaumarchaeota

Chordate

Chordates are animals possessing a notochord, a hollow dorsal nerve cord, pharyngeal slits, an endostyle, and a post-anal tail for at least some period of their life cycles. All Chordata are deuterostomes as in the embryo development stage the anus forms before the mouth.

Taxonomically, the phylum includes the subphyla Vertebrata, which includes fish, amphibians, reptiles, birds, and mammals; Tunicata, which includes salps and sea squirts; and Cephalochordata, comprising the lancelets.

Members of the phylum Chordata are bilaterally symmetric, deuterostome coelomates. Vetebrate chordates can have body plans organized via segmentation.

Hemichordata, which includes the acorn worms, has been presented as a fourth chordate subphylum, but it now is usually treated as a separate phylum. It, along with the phylum Echinodermata, which includes starfish, sea urchins, sea cucumbers, crinoids, are the chordates' closest relatives. Primitive chordates are known from at least as early as the Cambrian explosion.

Of the more than 65,000 living species of chordates, about half are bony fish of the class Osteichthyes. The world's largest and fastest animals, the blue whale and peregrine falcon respectively, are chordates, as are humans.

Overview of Affinities

Attempts to work out the evolutionary relationships of the chordates have produced several hypotheses. The current consensus is that chordates are monophyletic, meaning that the Chordata include all and only the descendants of a single common ancestor, which is itself a chordate, and that craniates' nearest relatives are cephalochordates.

All of the earliest chordate fossils have been found in the Early Cambrian Chengjiang fauna, and include two species that are regarded as fish, which implies that they are vertebrates. Because the fossil record of early chordates is poor, only molecular phylogenetics offers a reasonable prospect of dating their emergence. However, the use of molecular phylogenetics for dating evolutionary transitions is controversial.

It has also proved difficult to produce a detailed classification within the living chordates. Attempts to produce evolutionary "family trees" shows that many of the traditional classes are paraphyletic.

While this has been well known since the 19th century, an insistence on only monophyletic taxa has resulted in vertebrate classification being in a state of flux.

Origin of Name

Although the name Chordata is attributed to William Bateson (1885), it was already in prevalent use by 1880. Ernst Haeckel described a taxon comprising tunicates, cephalochordates, and vertebrates in 1866. Though he used the German vernacular form, it is allowed under the ICZN code because of its subsequent latinization.

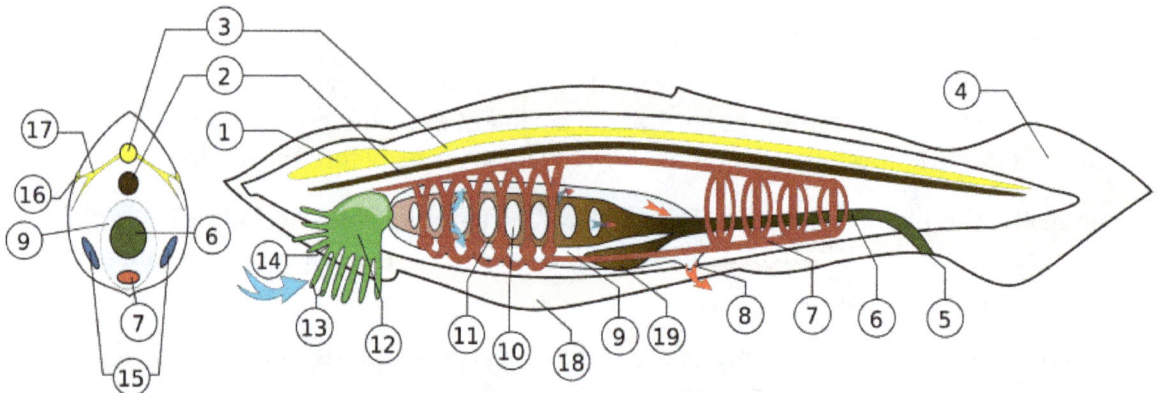

Anatomy of the cephalochordate *Amphioxus*. Bolded items are components of all chordates at some point in their lifetimes, and distinguish them from other phyla.

1 = bulge in spinal cord ("brain")

2 = notochord

3 = dorsal nerve cord

4 = post-anal tail

5 = anus

6 = digestive canal

7 = circulatory system

8 = atriopore

9 = space above pharynx

10 = pharyngeal slit (gill)

11 = pharynx

12 = vestibule

13 = oral cirri

14 = mouth opening

15 = gonads (ovary / testicle)

16 = light sensor

17 = nerves

18 = metapleural fold

19 = hepatic caecum (liver-like sack)

Definition

Chordates form a phylum of creatures that are based on a bilateral body plan, and is defined by having at some stage in their lives all of the following:

- A notochord, a fairly stiff rod of cartilage that extends along the inside of the body. Among the vertebrate sub-group of chordates the notochord develops into the spine, and in wholly aquatic species this helps the animal to swim by flexing its tail.

- A dorsal neural tube. In fish and other vertebrates, this develops into the spinal cord, the main communications trunk of the nervous system.

- Pharyngeal slits. The pharynx is the part of the throat immediately behind the mouth. In fish, the slits are modified to form gills, but in some other chordates they are part of a filter-feeding system that extracts particles of food from the water in which the animals live.

- Post-anal tail. A muscular tail that extends backwards behind the anus.

- An endostyle. This is a groove in the ventral wall of the pharynx. In filter-feeding species it produces mucus to gather food particles, which helps in transporting food to the esophagus. It also stores iodine, and may be a precursor of the vertebrate thyroid gland.

There are some soft constraints that separate Chordata from other biological lineages, but have not yet been made part of the formal definition.

- Pigmentation : Chordata are not capable of blue pigmentation, that is to say blue skin pigmentation. Blue skin pigmentation is not to be confused with the biological alteration of (for example) feathers in birds -- which appear blue via an evolved prism mechanism in the feather structure.

- All Chordata are deuterostomes. This means that, during the embryo development stage, the anus forms before the mouth.

There still is a lot of differential (DNA sequence based) comparison research going on that is trying to separate out the simplest forms of chordata. As around 90% of the species on Earth don't have a backbone like chordata.

As some lineages of the 90% of species that don't have a backbone or notochord may have lost these structures over time, this complicates the classification of chordata. Some Chordata lineages may only be found by DNA analysis, when there is no physical trace of any chordata like structures.

Subdivisions

Craniata (Vertebrata)

Craniate: Hagfish

Craniates, one of the three subdivisions of chordates, have distinct skulls - including hagfish, which have no vertebrae. Michael J. Benton comments, "craniates are characterized by their heads, just as chordates, or possibly all deuterostomes, are by their tails".

Most are vertebrates, in which the notochord is replaced by the spinal column, which consists of a series of bony or cartilaginous cylindrical vertebrae, generally with neural arches that protect the spinal cord and with projections that link the vertebrae. Hagfish have incomplete braincases and no vertebrae, and are therefore not regarded as vertebrates, but as members of the craniates, the group from which vertebrates are thought to have evolved. The position of lampreys is ambiguous. They have complete braincases and rudimentary vertebrae, and therefore may be regarded as vertebrates and true fish. However, molecular phylogenetics, which uses biochemical features to classify organisms, has produced both results that group them with vertebrates and others that group them with hagfish.

Tunicata (Tunicates, or Urochordates)

Comparison of Three Invertebrate Chordates

Most tunicates appear as adults in two major forms, both of which are soft-bodied filter-feeders that lack the standard features of chordates: "sea squirts" are sessile and consist mainly of water pumps and filter-feeding apparatus; salps float in mid-water, feeding on plankton, and have a two-generation cycle in which one generation is solitary and the next forms chain-like colonies. However, all tunicate larvae have the standard chordate features, including long, tadpole-like tails; they also have rudimentary brains, light sensors and tilt sensors. The third main group of tunicates, Appendicularia (also known as Larvacea) retain tadpole-like shapes and active swimming all their lives, and were for a long time regarded as larvae of sea squirts or salps. The etymology of the term Urochorda(ta) (Balfour 1881) is from the ancient (oura, "tail") + Latin chorda ("cord"), because the notochord is only found in the tail. The term Tunicata (Lamarck 1816) is recognised as having precedence and is now more commonly used.

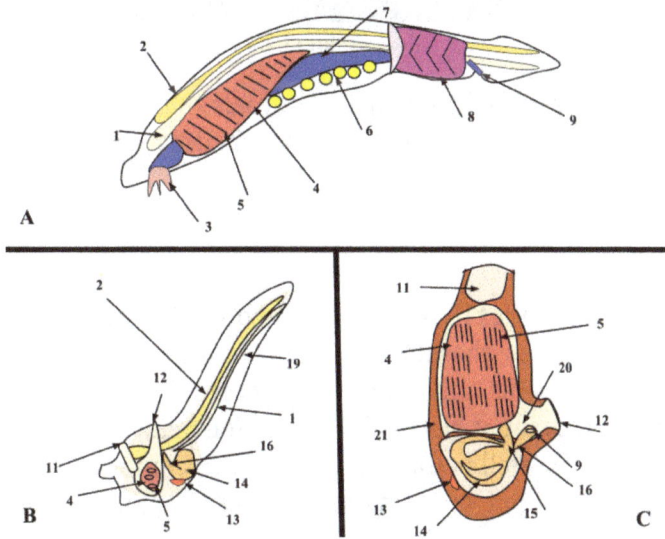

A. Lancelet, B. Larval tunicate, C. Adult tunicate

1. Notochord, 2. Nerve chord, 3. Buccal cirri, 4. Pharynx, 5. Gill slit, 6. Gonad, 7. Gut, 8. V-shaped muscles, 9. Anus, 10. Inhalant syphon, 11. Exhalant syphon, 12. Heart, 13. Stomach, 14. Esophagus, 15. Intestines, 16. Tail, 17. Atrium, 18. Tunic

Tunicates: sea squirts

Cephalochordata: Lancelets

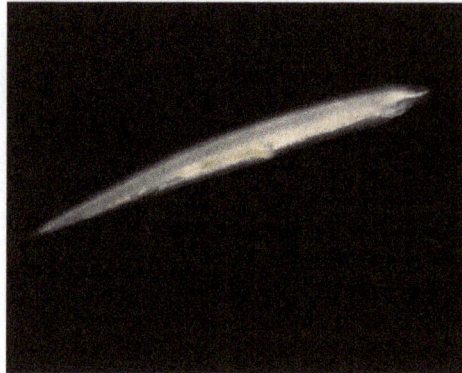

Cephalochordate: Lancelet

Cephalochordates are small, "vaguely fish-shaped" animals that lack brains, clearly defined heads and specialized sense organs. These burrowing filter-feeders comprise the earliest-branching chordate sub-phylum.

Origins

The majority of animals more complex than jellyfish and other Cnidarians are split into two groups, the protostomes and deuterostomes, the latter of which contains chordates. It seems very likely the 555 million-year-old *Kimberella* was a member of the protostomes. If so, this means the protostome and deuterostome lineages must have split some time before *Kimberella* appeared—at least 558 million years ago, and hence well before the start of the Cambrian 541 million years ago. The Ediacaran fossil *Ernietta*, from about 549 to 543 million years ago, may represent a deuterostome animal.

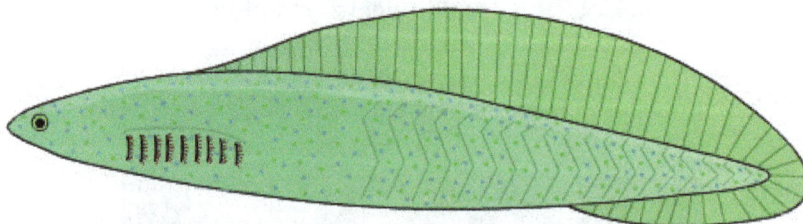

Haikouichthys, from about 518 million years ago in China, may be the earliest known fish.

Fossils of one major deuterostome group, the echinoderms (whose modern members include starfish, sea urchins and crinoids), are quite common from the start of the Cambrian, 542 million years ago. The Mid Cambrian fossil *Rhabdotubus johanssoni* has been interpreted as a pterobranch hemichordate. Opinions differ about whether the Chengjiang fauna fossil *Yunnanozoon*, from the earlier Cambrian, was a hemichordate or chordate. Another fossil, *Haikouella lanceolata*, also from the Chengjiang fauna, is interpreted as a chordate and possibly a craniate, as it shows signs of a heart, arteries, gill filaments, a tail, a neural chord with a brain at the front end, and possibly eyes—although it also had short tentacles round its mouth. *Haikouichthys* and *Myllokunmingia*, also from the Chengjiang fauna, are regarded as fish. *Pikaia*, discovered much earlier (1911) but from the Mid Cambrian Burgess Shale (505 Ma), is also regarded as a primitive chordate. On the other hand, fossils of early chordates are very rare, since non-vertebrate chordates have no bones or teeth, and only one has been reported for the rest of the Cambrian.

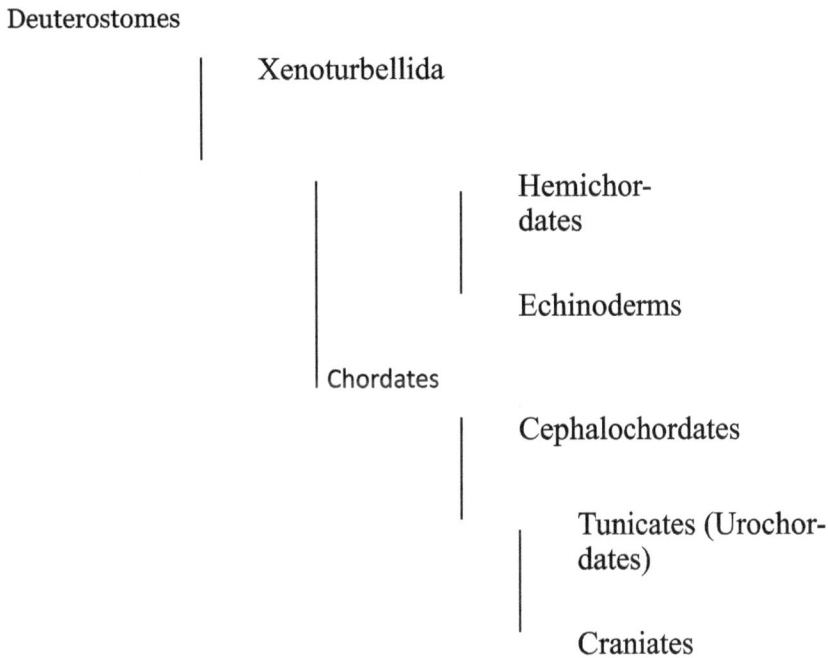

Deuterostomes

Xenoturbellida

Hemichor-
dates

Echinoderms

Chordates

Cephalochordates

Tunicates (Urochor-
dates)

Craniates

A consensus family tree of the chordates

The evolutionary relationships between the chordate groups and between chordates as a whole and their closest deuterostome relatives have been debated since 1890. Studies based on anatomical, embryological, and paleontological data have produced different "family trees". Some closely linked chordates and hemichordates, but that idea is now rejected. Combining such analyses with data from a small set of ribosome RNA genes eliminated some older ideas, but opened up the possibility that tunicates (urochordates) are "basal deuterostomes", surviving members of the group from which echinoderms, hemichordates and chordates evolved. Some researchers believe that, within the chordates, craniates are most closely related to cephalochordates, but there are also reasons for regarding tunicates (urochordates) as craniates' closest relatives. One other phylum, Xenoturbellida, has been thought for a while to be basal within the deuterostomes, closer to the original deuterostomes than to the chordates, echinoderms and hemichordates. But an article by Nakano et al. ("*Xenoturbella bocki* exhibits direct development with similarities to Acoelomorpha", *Nature Communications*, 2013; 4: 1537 DOI: 10.1038/ncomms2556) demonstrates that the Xenoturbellida are not deuterostomes.

Since early chordates have left a poor fossil record, attempts have been made to calculate the key dates in their evolution by molecular phylogenetics techniques—by analyzing biochemical differences, mainly in RNA. One such study suggested that deuterostomes arose before 900 million years ago and the earliest chordates around 896 million years ago. However, molecular estimates of dates often disagree with each other and with the fossil record, and their assumption that the molecular clock runs at a known constant rate has been challenged.

Classification

Taxonomy

A skeleton of the blue whale, the world's largest animal, outside the Long Marine Laboratory at the University of California, Santa Cruz

A peregrine falcon, the world's fastest animal

Traditionally, Cephalochordata and Craniata were grouped into the proposed clade "Euchordata", which would have been the sister group to Tunicata/Urochordata. More recently, Cephalochordata has been thought of as a sister group to the "Olfactores", which includes the craniates and tunicates. The matter is not yet settled.

The following schema is from the third edition of *Vertebrate Palaeontology*. The invertebrate chordate classes are from Fishes of the World. While it is structured so as to reflect evolutionary relationships (similar to a cladogram), it also retains the traditional ranks used in Linnaean taxonomy.

- Phylum Chordata
  - Subphylum Tunicata (Urochordata) – (tunicates; 3,000 species)
    - Class Ascidiacea (sea squirts)
    - Class Thaliacea (salps)
    - Class Appendicularia (larvaceans)
    - Class Sorberacea

- o Subphylum Cephalochordata (Acraniata) – (lancelets; 30 species)
  - Class Leptocardii (lancelets)
- o Subphylum Vertebrata (Craniata) (vertebrates – animals with backbones; 57,674 species)
  - Infraphylum incertae sedis (Cyclostomata)
    - Superclass 'Agnatha' paraphyletic (jawless vertebrates; 100+ species)
      - Class Myxini (hagfish; 65 species)
      - Class Petromyzontida or Hyperoartia (lampreys)
      - Class †Conodonta
      - Class †Pteraspidomorphi
      - Class †Thelodonti
      - Class †Anaspida
      - Class †Cephalaspidomorphi
  - Infraphylum Gnathostomata (jawed vertebrates)
    - Class †Placodermi (Paleozoic armoured forms; paraphyletic in relation to all other gnathostomes)
    - Class Chondrichthyes (cartilaginous fish; 900+ species)
    - Class †Acanthodii (Paleozoic "spiny sharks"; paraphyletic in relation to Chondrichthyes)
    - Superclass Osteichthyes (bony fish; 30,000+ species)
      - Class Actinopterygii (ray-finned fish; about 30,000 species)
      - Class Sarcopterygii (lobe-finned fish: 8 species)
    - Superclass Tetrapoda (four-limbed vertebrates; 28,000+ species) (The classification below follows Benton 2004, and uses a synthesis of rank-based Linnaean taxonomy and also reflects evolutionary relationships. Benton included the Superclass Tetrapoda in the Subclass Sarcopterygii in order to reflect the direct descent of tetrapods from lobe-finned fish, despite the former being assigned a higher taxonomic rank.)
      - Class Amphibia (amphibians; 7,000+ species)
      - Class Sauropsida (reptiles (including birds); 9,000+ species)
      - Class Synapsida (mammals; 5,700+ species)

Closest Nonchordate Relatives

Hemichordates

Hemichordates ("half (½) chordates") have some features similar to those of chordates: branchial openings that open into the pharynx and look rather like gill slits; stomochords, similar in composition to notochords, but running in a circle round the "collar", which is ahead of the mouth; and a dorsal nerve cord—but also a smaller ventral nerve cord.

There are two living groups of hemichordates. The solitary enteropneusts, commonly known as "acorn worms", have long proboscises and worm-like bodies with up to 200 branchial slits, are up to 2.5 metres (8.2 ft) long, and burrow though seafloor sediments. Pterobranchs are colonial animals, often less than 1 millimetre (0.039 in) long individually, whose dwellings are interconnected. Each filter feeds by means of a pair of branched tentacles, and has a short, shield-shaped proboscis. The extinct graptolites, colonial animals whose fossils look like tiny hacksaw blades, lived in tubes similar to those of pterobranchs.

Echinoderms

Echinoderms differ from chordates and their other relatives in three conspicuous ways: they possess bilateral symmetry only as larvae - in adulthood they have radial symmetry, meaning that their body pattern is shaped like a wheel; they have tube feet; and their bodies are supported by skeletons made of calcite, a material not used by chordates. Their hard, calcified shells keep their bodies well protected from the environment, and these skeletons enclose their bodies, but are also covered by thin skins. The feet are powered by another unique feature of echinoderms, a water vascular system of canals that also functions as a "lung" and surrounded by muscles that act as pumps. Crinoids look rather like flowers, and use their feather-like arms to filter food particles out of the water; most live anchored to rocks, but a few can move very slowly. Other echinoderms are mobile and take a variety of body shapes, for example starfish, sea urchins and sea cucumbers.

Autotomy

A white-headed dwarf gecko with tail lost due to autotomy.

Autotomy (from the Greek *auto-* "self-" and *tome* "severing") or self amputation is the behaviour whereby an animal sheds or discards one or more of its own appendages, usually as a self-defense mechanism to elude a predator's grasp or to distract the predator and thereby allow escape. The lost body part may be regenerated later.

Vertebrates

Reptiles and Amphibians

Some geckos, skinks, lizards, salamanders and tuatara that are captured by the tail will shed part of the tail structure and thus be able to flee. The detached tail will continue to wriggle, creating a deceptive sense of continued struggle and distracting the predator's attention from the fleeing prey animal. The animal can partially regenerate its tail, typically over a period of weeks. The new section will contain cartilage rather than regenerating vertebrae of bone, and the skin of the regenerated organ generally differs distinctly in colour and texture from its original appearance. The technical term for this ability to drop the tail is *caudal autotomy*. In most lizards that sacrifice the tail in this manner, breakage occurs only when the tail is grasped with sufficient force, but some animals, such as some species of geckos, can perform true autotomy, throwing off the tail when sufficiently stressed, such as when attacked by ants. Some such lizards, in which the tail is a major storage organ for accumulating reserves, will return to a dropped tail after the threat has passed, and will eat it to recover part of the sacrificed supplies. Conversely, some species have been observed to attack rivals and grab their tails, which they eat after their opponents flee.

Lizard tail autotomy

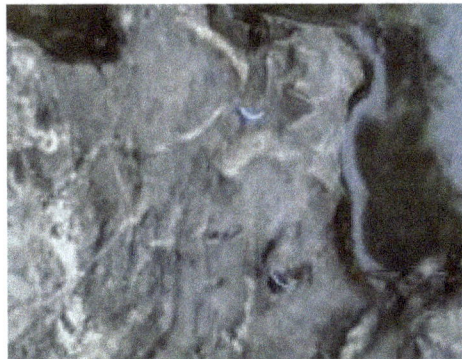
Discarded salamander tails exhibiting continued movement

Caudal autotomy in lizards takes two forms. In the first form, called intervertebral autotomy, the tail breaks between the vertebrae. The second form of caudal autotomy is intravertebral autotomy, in which there are zones of weakness, fracture planes across each vertebra in the mid-part of the tail. In this second type of autotomy the lizard contracts a muscle to fracture a vertebra, rather than break the tail between two vertebrae. Sphincter muscles in the tail then contract around the caudal artery to minimize bleeding.

Mammals

At least two species of African spiny mice, *Acomys kempi* and *Acomys percivali*, are capable of autotomic release of skin, e.g. upon being captured by a predator. They are the first mammals known to do so. They can completely regenerate the autotomically released or otherwise damaged skin tissue — regrowing hair follicles, skin, sweat glands, fur and cartilage with little or no scarring. It is believed that the corresponding regeneration genes could also function in humans.

Invertebrates

Over 200 species of invertebrates are capable of using autotomy as an avoidance or protective behaviour including,

Mollusca:

- land slugs *(Prophysaon)*
- sea snails *(Oxynoe panamensis)*
- octopuses

Arthropoda:

- crickets
- spiders
- crabs
- lobsters

These animals can voluntarily shed appendages when necessary for survival. Autotomy can occur in response to chemical, thermal and electrical stimulation, but is perhaps most frequently a response to mechanical stimulation during capture by a predator. Autotomy serves either to improve the chances of escape or to reduce further damage occurring to the remainder of the animal such as the spread of a chemical toxin after being stung.

Molluscs

Autotomy occurs in some species of octopus for survival and for reproduction: the specialized reproductive arm (the hectocotylus) detaches from the male during mating and remains within the female's mantle cavity.

Species of (land) slugs in the genus *Prophysaon* can self-amputate a portion of their tail. There is known autotomy of the tail of sea snail *Oxynoe panamensis* under persistent mechanical irritation.

Evisceration, the ejection of the internal organs of sea cucumbers when stressed, is also a form of autotomy, and they regenerate the organ(s) lost.

Some sea slugs exhibit autotomy. Both *Discodoris lilacina* and *Berthella martensi* will often drop their entire mantle skirt when handled, leading to Discodoris lilacina also being called *Discodoris fragilis*. The members of *Phyllodesmium* will drop a large number of their cerata each, on the tip having a large sticky gland that secretes a sticky substance.

Crustaceans

Autotomic stone crabs are used as a self-replenishing source of food by humans, particularly in Florida. Harvesting is accomplished by removing one or both claws from the live animal and returning it to the ocean where it can regrow the lost limb(s). However, under experimental conditions, but using commercially accepted techniques, 47% of stone crabs that had both claws removed died after declawing, and 28% of single claw amputees died; 76% of the casualties died within 24 hours of declawing. The occurrence of regenerated claws in the fishery harvest is low; one study indicates less than 10%, and a more recent study indicates only 13% have regenerated claws.

Post-harvest leg autotomy can be problematic in some crab and lobster fisheries, and often occurs if these crustaceans are exposed to freshwater or hypersaline water in the form of dried salt on sorting trays. The autotomy reflex in crustaceans has been proposed as an example of natural behaviour that raises questions concerning assertions on whether crustaceans can "feel pain", which may be based on definitions of "pain" that are flawed for lack of any falsifiable test, either to establish or deny the meaningfulness of the concept in this context.

Spiders

A fishing spider with two limbs missing

Under natural conditions, orb-weaving spiders (*Argiope* spp.) undergo autotomy if they are stung in a leg by wasps or bees. Under experimental conditions, when spiders are injected in the leg with bee or wasp venom, they shed this appendage. But, if they are injected with only saline, they rarely autotomize the leg, indicating it is not the physical injection or the ingress of fluid *per se* that causes autotomy. In addition, spiders injected with venom components which cause injected humans to report pain (serotonin, histamine, phospholipase A2 and melittin) autotomize the leg, but if the injections contain venom components which do not cause pain to humans, autotomy does not occur.

Bees

Sometimes when honey bees (genus *Apis*) sting a victim, the barbed stinger remains embedded. As the bee tears itself loose, the stinger takes with it the entire distal segment of the bee's abdomen, along with a nerve ganglion, various muscles, a venom sac, and the end of the bee's digestive tract. This massive abdominal rupture kills the bee. Although it is widely believed that a worker honey bee can sting only once, this is a partial misconception: although the stinger is barbed so that it lodges in the victim's skin, tearing loose from the bee's abdomen and leading to its death, this only happens if the skin of the victim is sufficiently thick, such as a mammal's. The sting of a queen honey bee has no barbs, however, and does not autotomize. All species of true honey bees have this form of stinger autotomy. No other stinging insect, including the yellowjacket wasp and the Mexican honey wasp, have the sting apparatus modified this way, though they may have barbed stings. Two wasp species that use sting autotomy as a defense mechanism are *Polybia rejecta* and *Synoeca surinama*.

The endophallus and cornua portions of the genitalia of male honey bees (drones) also autotomize during copulation, and form a mating plug, which must be removed by the genitalia of subsequent drones if they are also to mate with the same queen. The drones die within minutes of mating.

Bipedalism

Bipedalism is a form of terrestrial locomotion where an organism moves by means of its two rear limbs or legs. An animal or machine that usually moves in a bipedal manner is known as a biped, meaning "two feet" (from the Latin *bis* for "double" and *pes* for "foot"). Types of bipedal movement include walking, running, or hopping.

An ostrich, the fastest extant biped at 70 km/h[a]

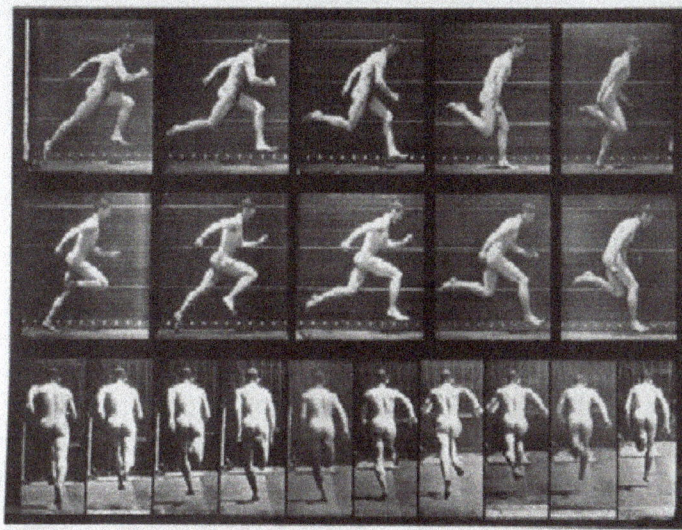

A Man Running - Eadweard Muybridge

Few modern species are habitual bipeds whose normal method of locomotion is two-legged. Within mammals, habitual bipedalism has evolved multiple times, with the macropods, kangaroo rats and mice, springhare, hopping mice, pangolins and homininan apes, as well as various other extinct groups evolving the trait independently. In the Triassic period some groups of archosaurs (a group that includes the ancestors of crocodiles) developed bipedalism; among their descendants the dinosaurs, all the early forms and many later groups were habitual or exclusive bipeds; the birds descended from one group of exclusively bipedal dinosaurs.

A larger number of modern species intermittently or briefly use a bipedal gait. Several non-archosaurian lizard species move bipedally when running, usually to escape from threats. Many primate and bear species will adopt a bipedal gait in order to reach food or explore their environment. Several arboreal primate species, such as gibbons and indriids, exclusively walk on two legs during the brief periods they spend on the ground. Many animals rear up on their hind legs whilst fighting or copulating. Some animals commonly stand on their hind legs, in order to reach food, to keep watch, to threaten a competitor or predator, or to pose in courtship, but do not move bipedally.

Definition

The word is derived from the Latin words *bi(s)* 'two' and *ped-* 'foot', as contrasted with quadruped 'four feet'.

Advantages

Limited and exclusive bipedalism can offer a species several advantages. Bipedalism raises the head; this allows a greater field of vision with improved detection of distant dangers or resources, access to deeper water for wading animals and allows the animals to reach higher food sources with their mouths. While upright, non-locomotory limbs become free for other uses, including manipulation (in primates and rodents), flight (in birds), digging (in giant pangolin), combat (in bears, great apes and the large monitor lizard) or camouflage (in certain species of octopus). The maximum bipedal speed appears less fast than the maximum speed of quadrupedal movement with a flexible backbone – both the ostrich and the red kangaroo can reach speeds of 70 km/h (43 mph), while the cheetah can exceed 100 km/h (62 mph). Bipedality in kangaroo rats has been hypothesized to improve locomotor performance, which could aid in escaping from predators.

Facultative and Obligate Bipedalism

Zoologists often label behaviors, including bipedalism, as "facultative" (i.e. optional) or "obligate" (the animal has no reasonable alternative). Even this distinction is not completely clear-cut — for example, humans other than infants normally walk and run in biped fashion, but almost all can crawl on hands and knees when necessary. There are even reports of humans who normally walk on all fours with their feet but not their knees on the ground, but these cases are a result of conditions such as Uner Tan syndrome — very rare genetic neurological disorders rather than normal behavior. Even if one ignores exceptions caused by some kind of injury or illness, there are many unclear cases, including the fact that "normal" humans can crawl on hands and knees. This article therefore avoids the terms "facultative" and "obligate", and focuses on the range of styles of locomotion *normally* used by various groups of animals.

Movement

There are a number of states of movement commonly associated with bipedalism.

1. Standing. Staying still on both legs. In most bipeds this is an active process, requiring constant adjustment of balance.

2. Walking. One foot in front of another, with at least one foot on the ground at any time.

3. Running. One foot in front of another, with periods where both feet are off the ground.

4. Jumping/hopping. Moving by a series of jumps with both feet moving together.

Bipedal Animals

The great majority of living terrestrial vertebrates are quadrupeds, with bipedalism exhibited by only a handful of living groups. Humans, gibbons and large birds walk by raising one foot at a time. On the other hand, most macropods, smaller birds, lemurs and bipedal rodents move by hopping on both legs simultaneously. Tree kangaroos are able to walk or hop, most commonly alternating feet when moving arboreally and hopping on both feet simultaneously when on the ground.

Amphibians

There are no known living or fossil bipedal amphibians.

Extant Reptiles

Many species of lizards become bipedal during high-speed, sprint locomotion, including the world's fastest lizard, the spiny-tailed iguana (genus *Ctenosaura*).

Early Reptiles and Lizards

The first known biped is the bolosaurid *Eudibamus* whose fossils date from 290 million years ago. Its long hindlegs, short forelegs, and distinctive joints all suggest bipedalism. The species was extinct before the dinosaurs appeared.

Archosaurs (Include Birds, Crocodiles, and Dinosaurs)

Birds

All birds are bipeds when on the ground, a feature inherited from their dinosaur ancestors.

Other Archosaurs

Bipedalism evolved more than once in archosaurs, the group that includes both dinosaurs and crocodilians. All dinosaurs are thought to be descended from a fully bipedal ancestor, perhaps similar to *Eoraptor*. Bipedal movement also re-evolved in a number of other dinosaur lineages such as the iguanodons. Some extinct members of the crocodilian line, a sister group to the dinosaurs and birds, also evolved bipedal forms - a crocodile relative from the triassic, *Effigia okeeffeae*, is thought to be bipedal. Pterosaurs were previously thought to have been bipedal, but recent trackways have all shown quadrupedal locomotion. Bipedalism also evolved independently among the dinosaurs. Dinosaurs diverged from their archosaur ancestors approximately 230 million years ago during the Middle to Late Triassic period, roughly 20 million years after the Permian-Triassic extinction event wiped out an estimated 95% of all life on Earth. Radiometric dating of fossils from the early dinosaur genus *Eoraptor* establishes its presence in the fossil record at this time. Paleontologists suspect *Eoraptor* resembles the common ancestor of all dinosaurs; if this is true, its traits suggest that the first dinosaurs were small, bipedal predators. The discovery of primitive, dinosaur-like ornithodirans such as *Marasuchus* and *Lagerpeton* in Argentinian Middle Triassic strata supports this view; analysis of recovered fossils suggests that these animals were indeed small, bipedal predators.

Mammals

A number of groups of extant mammals have independently evolved bipedalism as their main form of locomotion - for example humans, giant pangolins, the extinct giant ground sloths, numerous species of jumping rodents and macropods. Humans, as their bipedalism has been extensively studied, are documented in the next section. Macropods are believed to have evolved bipedal hopping only once in their evolution, at some time no later than 45 million years ago. Bipedal movement is less common among mammals, most of which are quadrupedal. All primates possess some bipedal ability, though most species primarily use quadrupedal locomotion on land. Primates aside, the macropods (kangaroos, wallabies and their relatives), kangaroo rats and mice, hopping mice and springhare move bipedally by hopping. Very few mammals other than primates commonly move bipedally by an alternating gait rather than hopping. Exceptions are the ground pangolin and in some circumstances the tree kangaroo.

Primates

Most bipedal animals move with their backs close to horizontal, using a long tail to balance the weight of their bodies. The primate version of bipedalism is unusual because the back is close to upright (completely upright in humans). Many primates can stand upright on their hind legs without any support. Chimpanzees, bonobos, gibbons and baboons exhibit forms of bipedalism. Injured chimpanzees and bonobos have been capable of sustained bipedalism. Geladas, although often quadrupedal, will move between adjacent feeding patches with a squatting, shuffling bipedal form of locomotion . Three captive primates, one macaque Natasha and two chimps, Oliver and Poko (chimpanzee), were found to move bipedally . Natasha switched to exclusive bipedalism after an illness, while Poko was discovered in captivity in a tall, narrow cage. Oliver reverted to knuckle-walking after developing arthritis. Non-human primates often use bipedal locomotion when carrying food.

The evolution of human bipedalism, began in primates about four million years ago, or as early as seven million years ago with *Sahelanthropus*. One hypothesis for human bipedalism is that it evolved as a result of differentially successful survival from carrying food to share with group members, although there are other hypotheses, as discussed below.

Limited Bipedalism

Limited Bipedalism in Mammals

Other mammals engage in limited, non-locomotory, bipedalism. A number of other animals, such as rats, raccoons, and beavers will squat on their hindlegs to manipulate some objects but revert to four limbs when moving (the beaver will move bipedally if transporting wood for their dams, as will the raccoon when holding food). Bears will fight in a bipedal stance to use their forelegs as weapons. A number of mammals will adopt a bipedal stance in specific situations such as for feeding or fighting. Ground squirrels and meerkats will stand on hind legs to survey their surroundings, but will not walk bipedally. Dogs (e.g. Faith) can stand or move on two legs if trained, or if birth defect or injury precludes quadrupedalism. The gerenuk antelope stands on its hind legs while eating from trees, as did the extinct giant ground sloth and chalicotheres. The spotted skunk will walk on its front legs when threatened, rearing up on its front legs while facing the attacker so that its anal glands, capable of spraying an offensive oil, face its attacker.

Limited Bipedalism in Non-Mammals

Bipedalism is unknown among the amphibians. Among the non-archosaur reptiles bipedalism is rare, but it is found in the 'reared-up' running of lizards such as agamids and monitor lizards. Many reptile species will also temporarily adopt bipedalism while fighting. One genus of basilisk lizard can run bipedally across the surface of water for some distance. Among arthropods, cockroaches are known to move bipedally at high speeds. Bipedalism is rarely found outside terrestrial animals, though at least two types of octopus walk bipedally on the sea floor using two of their arms, allowing the remaining arms to be used to camouflage the octopus as a mat of algae or a floating coconut.

There are at least twelve distinct hypotheses as to how and why bipedalism evolved in humans, and also some debate as to when. Bipedalism evolved well before the large human brain or the development of stone tools. Bipedal specializations are found in *Australopithecus* fossils from 4.2-3.9 million years ago, although *Sahelanthropus* may have walked on two legs as early as seven million years ago. Nonetheless, the evolution of bipedalism was accompanied by significant evolutions in the spine including the forward movement in position of the foramen magnum, where the spinal cord leaves the cranium. Recent evidence regarding modern human sexual dimorphism (physical differences between male and female) in the lumbar spine has been seen in pre-modern primates such as *Australopithecus africanus*. This dimorphism has been seen as an evolutionary adaptation of females to bear lumbar load better during pregnancy, an adaptation that non-bipedal primates would not need to make. Adapting bipedalism would have required less shoulder stability, which allowed the shoulder and other limbs to become more independent of each other and adapt for specific suspensory behaviors. In addition to the change in shoulder stability, changing locomotion would have increased the demand for shoulder mobility, which would have propelled the evolution of bipedalism forward. The different hypotheses are not necessarily mutually exclusive and a number of selective forces may have acted together to lead to human bipedalism. It is important to distinguish between adaptations for bipedalism and adaptations for running, which came later still.

Possible reasons for the evolution of human bipedalism include freeing the hands for tool use and carrying, sexual dimorphism in food gathering, changes in climate and habitat (from jungle to savanna) that favored a more elevated eye-position, and to reduce the amount of skin exposed to the tropical sun. It is possible that bipedalism provided a variety of benefits to the hominin species, and scientists have suggested multiple reasons for evolution of human bipedalism. There also is not only question of why were the earliest hominins partially bipedal but also why did hominins become more bipedal over time. For example, the postural feeding hypothesis (reaching for food/balancing) provides an explanation for how earliest hominins became for the benefit of reaching out for food in trees while the savannah-based theory describes how the late hominins that started to settle on the ground became increasingly bipedal.

Multiple Factors

Napier (1964) argued that it was very unlikely that single factor drove the evolution of Bipedalism. He stated *"It seems unlikely that any single factor was responsible for such a dramatic change in behaviour. In addition to the advantages of accruing from ability to carry objects - food or otherwise - the improvement of the visual range and the freeing of the hands for purposes of defence and offence must equally have played their part as catalysts."* Sigmon argued that chimpanzees demonstrate bipedalism in different contexts, and one single factor should be used to explain bipedalism. preadaptation for human bipedalism. Day (1986) emphasized three major pressures that drove evolution of bipedalism 1.food acquisition 2. predator avoidance 3. Reproductive success. Ko (2015) states there are two questions regarding bipedalism 1. Why were the earliest hominins partially bipedal 2. why did hominins become more bipedal over time. He argues that these questions can be answered with combination of prominent theories such as Savanna-based, Postural feeding, and Provisioning.

Savanna-based Theory

According to the savanna-based theory, hominines descended from the trees and adapted to life on the savanna by walking erect on two feet. The theory suggests that early hominids were forced to adapt to bipedal locomotion on the open savanna after they left the trees. This theory is closely related to the knuckle-walking hypothesis, which states that human ancestors used quadrupedal locomotion on the savanna, as evidenced by morphological characteristics found in *Australopithecus anamensis* and *Australopithecus afarensis* forelimbs, and that it is less parsimonious to assume that knuckle walking developed twice in genera Pan and Gorilla instead of evolving it once as synapomorphy for Pan and Gorilla before losing it in Australopithecus. The evolution of an orthograde posture would have been very helpful on a savanna as it would allow the ability to look over tall grasses in order to watch out for predators, or terrestrially hunt and sneak up on prey. It was also suggested in P.E. Wheeler's "The evolution of bipedality and loss of functional body hair in hominids", that a possible advantage of bipedalism in the savanna was reducing the amount of surface area of the body exposed to the sun, helping regulate body temperature. In fact, Elizabeth Vrba's turnover pulse hypothesis supports the savanna-based theory by explaining the shrinking of forested areas due to global warming and cooling, which forced animals out into the open grasslands and caused the need for hominids to acquire bipedality.

Rather, the bipedal adaptation hominines had already achieved was used in the savanna. The fossil record shows that early bipedal hominines were still adapted to climbing trees at the time they were also walking upright. It is possible that Bipedalism evolved in the trees, and was later applied to the Savannah as a vestigial trait. Humans and orangutans are both unique to a bipedal reactive adaptation when climbing on thin branches, in which they have increased hip and knee extension in relation to the diameter of the branch, which can increase an arboreal feeding range and can be attributed to a convergent evolution of bipedalism evolving in arboreal environments. Hominine fossils found in dry grassland environments led anthropologists to believe hominines lived, slept, walked upright, and died only in those environments because no hominine fossils were found in forested areas. However, fossilization is a rare occurrence—the conditions must be just right in order for an organism that dies to become fossilized for somebody to find later, which is also a rare occurrence. The fact that no hominine fossils were found in forests does not ultimately lead to the conclusion that no hominines ever died there. The convenience of the savanna-based theory caused this point to be overlooked for over a hundred years.

Some of the fossils found actually showed that there was still an adaptation to arboreal life. For example, Lucy, the famous *Australopithecus afarensis*, found in Hadar in Ethiopia, which may have been forested at the time of Lucy's death, had curved fingers that would still give her the ability to grasp tree branches, but she walked bipedally. "Little Foot," the collection of *Australopithecus africanus* foot bones, has a divergent big toe as well as the ankle strength to walk upright. "Little Foot" could grasp things using his feet like an ape, perhaps tree branches, and he was bipedal. Ancient pollen found in the soil in the locations in which these fossils were found suggest that the area used to be much more wet and covered in thick vegetation and has only recently become the arid desert it is now.

Traveling Efficiency Hypothesis

An alternative explanation is the mixture of savanna and scattered forests increased terrestrial travel by proto-humans between clusters of trees, and bipedalism offered greater efficiency for long-distance travel between these clusters than quadrupedalism. In an experiment monitoring chimpanzee metabolic rate via oxygen consumption, it was found that the quadrupedal and bipedal energy costs were very similar, implying that this transition in early ape-like ancestors would have not have been very difficult or energetically costing. This increased travel efficiency is likely to have been selected for as it assisted the wide dispersal of early hominids across the Savannah to create start populations.

Postural Feeding Hypothesis

The postural feeding hypothesis has been recently supported by Dr. Kevin Hunt, a professor at Indiana University. This hypothesis asserts that chimpanzees were only bipedal when they eat. While on the ground, they would reach up for fruit hanging from small trees and while in trees, bipedalism was used to reach up to grab for an overhead branch. These bipedal movements may have evolved into regular habits because they were so convenient in obtaining food. Also, Hunt's hypotheses states that these movements coevolved with chimpanzee arm-hanging, as this movement was very effective and efficient in harvesting food. When analyzing fossil anatomy, *Australopithecus afarensis* has very similar features of the hand and shoulder to the chimpanzee, which indicates hanging arms. Also, the *Australopithecus* hip and hind limb very clearly indicate bipedalism, but these fossils also indicate very inefficient locomotive movement when compared to humans. For this reason, Hunt argues that bipedalism evolved more as a terrestrial feeding posture than as a walking posture.

A similar study conducted by Thorpe et al. looked at how the most arboreal great ape, the orangutan, held onto supporting branches in order to navigate branches that were too flexible or unstable otherwise. They found that in more than 75% of locomotive instances the orangutans used their hands to stabilize themselves while they navigated thinner branches. They hypothesized that increased fragmentation of forests where A. afarensis as well as other ancestors of modern humans and other apes resided could have contributed to this increase of bipedalism in order to navigate the diminishing forests. Their findings also shed light on a couple of discrepancies observed in the anatomy of A. afarensis, such as the ankle joint, which allowed it to "wobble" and long, highly flexible forelimbs. The idea that bipedalism started from walking in trees explains both the increased flexibility in the ankle as well as the long limbs which would be used to grab hold of branches.

Provisioning Model

One theory on the origin of bipedalism is the behavioral model presented by C. Owen Lovejoy, known as "male provisioning". Lovejoy theorizes that the evolution of bipedalism was linked to monogamy. In the face of long inter-birth intervals and low reproductive rates typical of the apes, early hominids engaged in pair-bonding that enabled greater parental effort directed towards rearing offspring. Lovejoy proposes that male provisioning of food would improve the offspring survivorship and increase the pair's reproductive rate. Thus the male would leave his mate and offspring to search for food and return carrying the food in his arms walking on his legs. This model is supported by the reduction ("feminization") of the male canine teeth in early hominids such as *Sahelanthropus tchadensis* and *Ardipithecus ramidus*, which along with low body size dimorphism in *Ardipithecus* and *Australopithecus*, suggests a reduction in inter-male antagonistic behavior in early hominids. In addition, this model is supported by a number of modern human traits associated with concealed ovulation (permanently enlarged breasts, lack of sexual swelling) and low sperm competition (moderate sized testes, low sperm mid-piece volume) that argues against recent adaptation to a polygynous reproductive system.

However, this model has generated some controversy, as others have argued that early bipedal hominids were instead polygynous. Among most monogamous primates, males and females are about the same size. That is sexual dimorphism is minimal, and other studies have suggested that Australopithecus afarensis males were nearly twice the weight of females. However, Lovejoy's model posits that the larger range a provisioning male would have to cover (to avoid competing with the female for resources she could attain herself) would select for increased male body size to limit predation risk. Furthermore, as the species became more bipedal, specialized feet would prevent the infant from conveniently clinging to the mother - hampering the mother's freedom and thus make her and her offspring more dependent on resources collected by others. Modern monogamous primates such as gibbons tend to be also territorial, but fossil evidence indicates that *Australopithecus afarensis* lived in large groups. However, while both gibbons and hominids have reduced canine sexual dimorphism, female gibbons enlarge ('masculinize') their canines so they can actively share in the defense of their home territory. Instead, the reduction of the male hominid canine is consistent with reduced inter-male aggression in a group living primate.

Early Bipedalism in Homininae Model

Recent studies of 4.4 million years old *Ardipithecus ramidus* suggest bipedalism, it is thus possible that bipedalism evolved very early in homininae and was reduced in chimpanzee and gorilla when they became more specialized. According to Richard Dawkins in his book "The Ancestor's Tale", chimps and bonobos are descended from *Australopithecus* gracile type species while gorillas are descended from Paranthropus. These apes may have once been bipedal, but then lost this ability when they were forced back into an arboreal habitat, presumably by those australopithecines who eventually became us. Early homininaes such as *Ardipithecus ramidus* may have possessed an arboreal type of bipedalism that later independently evolved towards knuckle-walking in chimpanzees and gorillas and towards efficient walking and running in modern humans. It is also proposed that one cause of Neanderthal extinction was a less efficient running.

Warning Display (Aposematic) Model

Joseph Jordania from the University of Melbourne recently (2011) suggested that bipedalism was one of the central elements of the general defense strategy of early hominids, based on aposematism, or warning display and intimidation of potential predators and competitors with exaggerated visual and audio signals. According to this model, hominids were trying to stay as visible and as loud as possible all the time. Several morphological and behavioral developments were employed to achieve this goal: upright bipedal posture, longer legs, long tightly coiled hair on the top of the head, body painting, threatening synchronous body movements, loud voice and extremely loud rhythmic singing/stomping/drumming on external subjects. Slow locomotion and strong body odor (both characteristic for hominids and humans) are other features often employed by aposematic species to advertise their non-profitability for potential predators.

Other Behavioural Models

There are a variety of ideas which promote a specific change in behaviour as the key driver for the evolution of hominid bipedalism. For example, Wescott (1967) and later Jablonski & Chaplin

(1993) suggest that bipedal threat displays could have been the transitional behaviour which led to some groups of apes beginning to adopt bipedal postures more often. Others (*e.g.* Dart 1925) have offered the idea that the need for more vigilance against predators could have provided the initial motivation. Dawkins (*e.g.* 2004) has argued that it could have begun as a kind of fashion that just caught on and then escalated through sexual selection. And it has even been suggested (*e.g.* Tanner 1981:165) that male phallic display could have been the initial incentive, as well as increased sexual signaling in upright female posture.

Thermoregulatory Model

The thermoregulatory model explaining the origin of bipedalism is one of the simplest theories so far advanced, but it is a viable explanation. Dr. Peter Wheeler, a professor of evolutionary biology, proposes that bipedalism raises the amount of body surface area higher above the ground which results in a reduction in heat gain and helps heat dissipation. When a hominid is higher above the ground, the organism accesses more favorable wind speeds and temperatures. During heat seasons, greater wind flow results in a higher heat loss, which makes the organism more comfortable. Also, Wheeler explains that a vertical posture minimizes the direct exposure to the sun whereas quadrupedalism exposes more of the body to direct exposure. Analysis and interpretations of Ardipithecus reveal that this hypothesis needs modification to consider that the forest and woodland environmental preadaptation of early-stage hominid bipedalism preceded further refinement of bipedalism by the pressure of natural selection. This then allowed for the more efficient exploitation of the hotter conditions ecological niche, rather than the hotter conditions being hypothetically bipedalism's initial stimulus. A feedback mechanism from the advantages of bipedality in hot and open habitats would then in turn make a forest preadaptation solidify as a permanent state.

Carrying Models

Charles Darwin wrote that "Man could not have attained his present dominant position in the world without the use of his hands, which are so admirably adapted to the act of obedience of his will" Darwin (1871:52) and many models on bipedal origins are based on this line of thought. Gordon Hewes (1961) suggested that the carrying of meat "over considerable distances" (Hewes 1961:689) was the key factor. Isaac (1978) and Sinclair et al. (1986) offered modifications of this idea as indeed did Lovejoy (1981) with his 'provisioning model' described above. Others, such as Nancy Tanner (1981) have suggested that infant carrying was key, whilst others have suggested stone tools and weapons drove the change. This stone tools theory is very unlikely, as though ancient humans were known to hunt, the discovery of tools was not discovered for thousands of years after the origin of bipedalism, temporally preventing it from being a driving force of evolution. (Wooden tools and spears fossilize poorly and therefore it's difficult to make a judgement about their potential usage.)

Wading Models

The observation that large Primates, including especially the great apes, that predominantly move quadrupedally on dry land, tend to switch to bipedal locomotion in waist deep water, has led to the idea that the origin of human bipedalism may have been influenced by waterside environments. This idea, labelled "The Wading Hypothesis", was originally promoted by Elaine Morgan, as part of the aquatic ape hypothesis, who cited bipedalism among a cluster of other human traits unique among primates, including voluntary control of breathing, hairlessness and subcutaneous fat. She argued that wading, swimming and diving through water offer better explanations for these traits than more conventional theories. The "aquatic ape hypothesis", as originally formulated, has not been accepted or considered a serious theory within the anthropological scholarly community. Others, however, have sought to promote wading as a factor in the origin of human bipedalism without referring to further ("aquatic ape" related) factors. Since 2000 Carsten Niemitz has published a series of papers and a book on a variant of the wading hypothesis, which he calls The Amphibian Generalist Theory. ("Amphibische Generalistentheorie").

Other theories have been proposed that suggest wading and the exploitation of aquatic food sources (providing essential nutrients for human brain evolution or critical fallback foods) may have exerted evolutionary pressures on human ancestors promoting adaptations which later assisted full-time bipedalism. It has also been thought that consistent water-based food sources had developed early hominid dependency and facilitated dispersal along seas and rivers.

Physiology

Bipedal movement occurs in a number of ways, and requires many mechanical and neurological adaptations. Some of these are described below.

Biomechanics

Standing

Energy-efficient means of standing bipedally involve constant adjustment of balance, and of course these must avoid overcorrection. The difficulties associated with simple standing in upright humans are highlighted by the greatly increased risk of falling present in the elderly, even with minimal reductions in control system effectiveness.

Shoulder Stability

Shoulder stability would decrease with the evolution of bipedalism. Shoulder mobility would increase because the need for a stable shoulder is only present in arboreal habitats. Shoulder mobility would support suspensory locomotion behaviors which are present in human bipedalism. The forelimbs are freed from weight bearing capabilities which makes the shoulder a place of evidence for the evolution of bipedalism.

Walking

Walking is characterized by an "inverted pendulum" movement in which the center of gravity vaults over a stiff leg with each step. Force plates can be used to quantify the whole-body kinetic & potential energy, with walking displaying an out-of-phase relationship indicating exchange between the two. Interestingly, this model applies to all walking organisms regardless of the number of legs, and thus bipedal locomotion does not differ in terms of whole-body kinetics.

In humans, walking is composed of several separate processes:

- Vaulting over a stiff stance leg
- Passive ballistic movement of the swing leg
- A short 'push' from the ankle prior to toe-off, propelling the swing leg
- Rotation of the hips about the axis of the spine, to increase stride length
- Rotation of the hips about the horizontal axis to improve balance during stance

Running

Running is characterized by a spring-mass movement. Kinetic and potential energy are in phase, and the energy is stored & released from a spring-like limb during foot contact. Again, the whole-body kinetics are similar to animals with more limbs.

Musculature

Bipedalism requires strong leg muscles, particularly in the thighs. Contrast in domesticated poultry the well muscled legs, against the small and bony wings. Likewise in humans, the quadriceps and hamstring muscles of the thigh are both so crucial to bipedal activities that each alone is much larger than the well-developed biceps of the arms.

Respiration

A biped has the ability to breathe while running, without strong coupling to stride cycle. Humans usually take a breath every other stride when their aerobic system is functioning. During a sprint the anaerobic system kicks in and breathing slows until the anaerobic system can no longer sustain a sprint.

Bipedal Robots

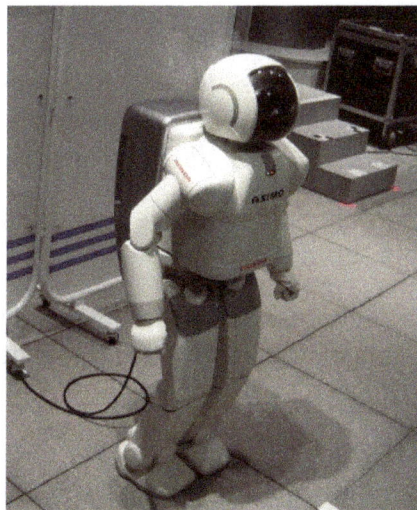

ASIMO - a bipedal robot

For nearly the whole of the 20th century, bipedal robots were very difficult to construct and robot locomotion involved only wheels, treads, or multiple legs. Recent cheap and compact computing power has made two-legged robots more feasible. Some notable biped robots are ASIMO, HUBO, MABEL and QRIO. Recently, spurred by the success of creating a fully passive, un-powered biped-al walking robot, those working on such machines have begun using principles gleaned from the study of human and animal locomotion, which often relies on passive mechanisms to minimize power consumption.

Alpha (Ethology)

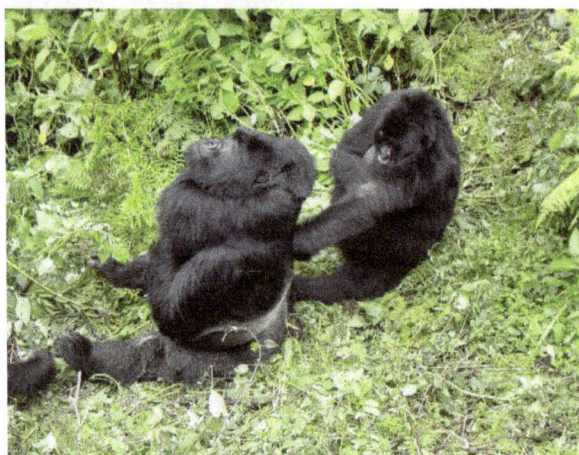

Male "silverback" gorillas are usually the alpha animal and may receive preferential treatment such as being groomed by other group members

In studies of social animals, the highest ranking individual is sometimes designated as the alpha. Males, females, or both, can be alphas, depending on the species. Where one male and one female fulfill this role together, they are sometimes referred to as the alpha pair. Other animals in the same social group may exhibit deference or other species-specific subordinate behaviours towards the alpha or alphas.

Alpha animals usually gain preferential access to food and other desirable items or activities, though the extent of this varies widely between species. Male or female alphas may gain preferential access to sex or mates; in some species, only alphas or an alpha pair reproduce.

Alphas may achieve their status by superior physical strength and aggression, or through social efforts and building alliances within the group.

The individual with alpha status sometimes changes, often through a fight between the dominant and a subordinate animal. Such fights may or may not be to the death, depending on the animal.

Beta and Omega

Social animals in a hierarchic community are sometimes assigned ranks in ethology studies.

Beta animals often act as second-in-command to the reigning alpha or alphas and will act as new alpha animals if an alpha dies or is otherwise no longer considered an alpha. In some species of birds, males pair up in twos when courting, the beta male aiding the alpha male. In wolves, the beta male is not generally allowed to mate with the females, but if the old alpha is removed or dies, he takes over the alpha's females, becoming the new alpha. It has been found that the social context of the animals has a significant impact on courtship behavior and the overall reproductive success of that animal.

Omega (usually rendered ω) is an antonym used to refer to the lowest caste of the hierarchical society. Omega animals are subordinate to all others in the community, and are expected by others in the group to remain submissive to everyone. Omega animals may also be used as communal scapegoats or outlets for frustration, or given the lowest priority when distributing food.

Examples

Primates

Common chimpanzees show deference to the alpha of the community by ritualized gestures such as bowing, allowing the alpha to walk first in a procession, or standing aside when the alpha challenges.

Gorillas use intimidation to establish and maintain alpha position. A study conducted regarding the reproductive behavior of male mountain gorillas (*Gorilla beringei beringei*) found further evidence that dominant males are favored to father offspring, even when there is a greater number of males in a notably enlarged group size. The study also concluded that mating access dropped off less steeply with status; alpha, beta, and gamma showing more similar mating success, compared to what had been previously thought.

A study on the association of alpha males and females during the non-breeding season in wild Capuchin monkeys (Cebus apella nigritus) examined whether alpha males are the preferred mate for females and, secondly, whether female-alpha status and relationship to the alpha-male can be explained through the individual characteristics and or social network of the female. The results indicated that alpha male Capuchin are the preferred mate for adult females. However, only the alpha females had strong interactions with the alpha males by virtue of a dominance hierarchy among the females in which only the most dominant and strong females were able to interact with the alpha male.

Canines

In the past, the prevailing view on gray wolf packs was that they consisted of individuals vying with each other for dominance, with dominant gray wolves being referred to as the "alpha" male and female, and the subordinates as "beta" and "omega" wolves. This terminology was first used in 1947 by Rudolf Schenkel of the University of Basel, who based his findings on researching the behavior of captive gray wolves. This view on gray wolf pack dynamics was later popularized by L. David Mech in his 1970 book *The Wolf*. He formally disavowed this terminology in 1999, explaining that it was heavily based on the behavior of captive packs consisting of unrelated individuals, an error reflecting the once prevailing view that wild pack formation occurred in winter among independent gray wolves. Later research on wild gray wolves revealed that the pack is usually a family consisting of a breeding pair and its offspring of the previous 1–3 years.

In some other wild canids, the alpha male may not have exclusive access to the alpha female; moreover, other pack members as in the African wild dog (*Lycaon pictus*) may guard the maternity den used by the alpha female.

References:

- Berg, Linda R. (2 March 2007). Introductory Botany: Plants, People, and the Environment (2 ed.). Cengage Learning. p. 15. ISBN 9780534466695. Retrieved 2012-07-23.

- Mauseth, James D. (2012). Botany : An Introduction to Plant Biology (5th ed.). Sudbury, MA: Jones and Bartlett Learning. ISBN 978-1-4496-6580-7. p. 489

- Crandall-Stotler, Barbara; Stotler, Raymond E. (2000). "Morphology and classification of the Marchantiophyta". In A. Jonathan Shaw & Bernard Goffinet (Eds.). Bryophyte Biology. Cambridge: Cambridge University Press. p. 21. ISBN 0-521-66097-1.

- Cree, A. (2002). Tuatara. In: Halliday, Tim and Adler, Kraig (eds.), The New Encyclopedia Of Reptiles and Amphibians. Oxford University Press, Oxford, pp. 210–211. ISBN 0-19-852507-9

- McDonnel, R.J., Paine, T.D. and Gormally, M.J., (2009). Slugs: A Guide to the Invasive and Native Fauna of California. 21 pp., ISBN 978-1-60107-564-2, page 9.

- Davies, S.J.J.F. (2003). "Birds I Tinamous and Ratites to Hoatzins". In Hutchins, Michael. Grzimek's Animal Life Encyclopedia. 8 (2 ed.). Farmington Hills, MI: Gale Group. pp. 99–101. ISBN 0-7876-5784-0.

- McHenry, H.M (2009). "Human Evolution". In Michael Ruse & Joseph Travis. Evolution: The First Four Billion Years. Cambridge, Massachusetts: The Belknap Press of Harvard University Press. p. 263. ISBN 978-0-674-03175-3.

- Cunnane, Stephen C (2005). Survival of the fattest: the key to human brain evolution. World Scientific Publishing Company. pp. 259. ISBN 981-256-191-9.

- de Waal, Frans (2007) [1982]. Chimpanzee Politics: Power and Sex Among Apes (25th Anniversary ed.). Baltimore, MD: JHU Press. ISBN 978-0-8018-8656-0. Retrieved 13 July 2011.

- Stewart, D. (2006-08-01). "A Bird Like No Other". National Wildlife. National Wildlife Federation. Archived from the original on 2012-02-09. Retrieved 2014-05-30.

- "Bipedality in chimpanzee (Pan troglodytes) and bonobo (Pan paniscus): Testing hypotheses on the evolution of bipedalism". .interscience.wiley.com. 2002-05-09. Retrieved 2013-04-30.

- "The 2006 Stock Assessment Update for the Stone Crab, Menippe spp., Fishery in Florida". Florida Fish and Wildlife Conservation Commission. Retrieved 23 September 2012.

- Tiddi, Barbara. "Social relationships between adult females and the alpha male in wild tufted Capuchin monkeys". American Journal of Primatology. Retrieved 7 April 2012.

- Wilson and colleagues (6 October 2009). "Seizing the Opportunity: Subordinate Male Fowl Respond Rapidly to Variation in Social Context. Ethology". Retrieved 4 August 2012.

Permissions

Index

www.ingramcontent.com/pod-product-compliance
Lightning Source LLC
Chambersburg PA
CBHW061245190326
41458CB00011B/3590